Ancient Egyptian Mathematical Papyri

Third Edition

Dr. Maher Y. Shawer

Originally printed and distributed in the United States by Createspace.com.

Third edition published by Kindle Direct Publishing
Copyright © 2021 by Mayer Y. Shawer

First Edition:
Copyright © 2017 by Maher Y. Shawer.

Library of Congress Cataloging in Publication Data
Library of Congress Control Number: 2017915237
Shawer, Maher Y.
Ancient egyptian mathematical papyri.
Originally published : United States of America
Includes bibliographical references and index
Third Edition
ISBN 13:9798453375400

*Hopefully this book will be an
encouragement for
the reader to search further and explore the
mathematical work of the ancient Egyptians.
It may be that this new and additional
information will surprise you.*

Maher Y. Shawer

CONTENTS

PREFACE

It was noted in the first printing that the history of Egypt is well known in the western world; the pyramids and other Egyptian artifacts have been studied and admired for hundreds of years. Unfortunately, little is known of the science and mathematics of ancient Egypt. Very little knowledge has been passed down about the 4000 years of ancient Egyptian mathematical history.

Most scholars of more modern times have concluded from a few mathematical papyri that ancient Egyptian mathematics was primitive, at best. Others have agreed with Neugebauer when he said, "Ancient science was a product of a very few men, and these few men happened not to be Egyptian" (Neugebauer, 1969, p. 41). Gillings quoted a historian as saying, "The ancient Egyptians lacked a scientific attitude of mind" (Gillings, 1981, p. 232). Struik wrote that, "All available texts point to an Egyptian mathematics of rather primitive standards" (Struik, 1948, p. 23). Jourdain wrote, "They did not reach the conception of scientific mathematics" (Jourdain, 1956, p. 12).

History seems to be silent regarding Greek scientists and mathematicians traveling to Egypt for their educations. However, it has been documented that Egypt was supreme in the leadership of civilization, and students from all parts flocked to that land seeking admission into her mysteries or wisdom system. Besides those born in Egypt, students also came from Ionia and the Islands of the Aegean to visit Egypt for their education. As regards the visiting of Greek mathematicians to Egypt for the purpose of their education, the following students are mentioned simply to establish the fact that Egypt was regarded as the educational center of the ancient world but, like the Jews, the Greeks also visited Egypt and received their education from the priests. Whether born there or visiting there, most of their important work was done in Egypt. Some of these scientists and mathematicians are: Thales, Pythagoras, Aristotle, Theodorus of Cyrene, Aristaeus the Elder, Thrasyllus of Mendes, Heron of Alexandria, and Euclid, just to name a few. (James, 1954, pp. 43-45.)

It is inconceivable to this author how scholars can come to these conclusions based on just a few papyri from over 4000 years ago without considering the mathematical propositions and theorems the ancient Egyptians used to solve the problems. Their conclusions are like an ant saying an elephant consists of only one toe. This attitude is even more surprising considering the grandeur and immensity of the pyramids, the government and economics of a country of nearly a thousand square miles, the design of an extensive canal system and sections of huge granaries, and all other proper organizations of a civilization that has existed successfully for centuries longer than that of any other nation in recorded history. Science and mathematics must have played an important role to a civilization that accomplished all of these achievements.

The ancient Egyptians developed three systems of writing: Hieroglyphic, Hieratic, and Demotic. In the science of mathematics, we have a total of 11 written papyri which survived to this day, and some historians of mathematics believe they cover the entire 4000 years. Six of these papyri were written in Hieratic and five were written in Demotic.

Most of the Hieratic Papyri were well researched, but the Demotic Papyri were not so well researched. There are some differences. In the Demotic Papyri, the calculations of the operations indicated in the procedure are no longer written within the text. Moreover, the multiplication procedures in the multiplication tables are different from the Hieratic Papyri. The problems also vary, such as the Mesopotamian problems of a "pole leaning against the wall." The Pythagorean Theory was used in solving these problems. The majority of them contain problems and solutions to these problems followed by a check to see that the solutions are correct.

Some of these papyri consist of only a single page and none appear to be treatises of mathematics, but a collection of mathematical exercises and examples. Scholars consider Ahmes' papyrus "a lesson book for a school boy." (Newton, 1956, p. 170.) It would be a mistake to say that the mathematics in these papyri represent all of the ancient Egyptian mathematics. However, some problems in these papyri are of a practical nature and some are of a mathematical nature. When you examine the conditions laid down and all the numbers involved in the practical problems, you will see that they are more like theoretical problems put in a concrete form. There is a scientific quality in their mathematics.

This book is not another history of mathematics. It is a book to explain all the mathematical concepts which the scribes used to solve the problems in all the ancient Egyptian mathematical papyri, both Hieratic and Demotic. The explanation will be in modern-day mathematics. It takes descriptions from the papyri, written by the scribes, of the problems and how they were solved. A similarity between the Hieratic and Demotic Mathematical Papyri are that both have used word problem texts and table texts. In the word problems, the problems were stated and data was given, followed by questions asking what was to be determined. The procedure to solve the problem is spelled out, step by step. Each instruction was comprised of arithmetic operations such as addition, subtraction, multiplication, division, etc. The last step was the solution, verified using the calculation in the instructions and, sometimes, accompanied by a drawing. The ancient Egyptians did not write the exact mathematical rules, formulae, theorems, or propositions that were used to arrive at their solutions because, at that time, there was a smaller store of mathematical facts and skills. It was up to this author to study and deduce which theorems, propositions, methods, and rules were used to solve the problems. Some of these methods are not used in modern mathematics books, but some methods are, and also are used by modern computers. The methods used in the papyrus are not in the textbooks of mathematics but are covered in this edition. This edition came with newer information from the previous edition. We will explain most of the methods used to solve more advanced problems and cover omissions not in the modern textbooks.

Researching ancient Egyptian mathematical texts has been difficult and confusing because of the fragmented condition of the papyri and some parts were lost. Many numerical mistakes and some drawing illustrations do not match the solutions. The reason is that the scribes mentally solved the problems and wrote their solutions. Their notes were also very brief. The mistakes are very easy to notice and will be pointed out to you. In this edition, these mistakes will be corrected and you will be very interested in knowing what the ancient Egyptian knew and his methods used to solve the problems. Some of the scholars overlooked the mathematical concepts, the theorems, and the propositions which the Egyptian scribe used to solve these problems.

A maxim of teaching mathematics is that the students are taught the propositions and theorems of mathematics which some mathematicians developed, and then are given problems to solve based on using the propositions and theorems. With this in mind, the author will explore some of the problems in various papyri and the mathematical concepts which the ancient Egyptians used, as well as the theorems and propositions that were used to solve these problems and prove them to be correct.

In this book, we will discuss the mathematical knowledge which is given or that which is omitted in the writings from the Hieratic and the Demotic Papyri. To make it easy to understand the ancient Egyptian mathematics, we translated the solutions of the scribes into modern mathematical format. We will also show that most of the solutions that the scribe wrote down followed a general mathematical formula, theorem, or proposition. We will match the solutions of the scribes with the concepts that were actually used to solve the problems. It is hoped that this clears the confusion and misunderstanding of the treasure left to us in mathematics by the ancient Egyptian.

The original research for this book covered over fifteen years that included much research in the United States at the University of Pennsylvania in Philadelphia and the Library of Congress in Washington, D.C., and travel to several continents. The author went to the British Museum where he originally gathered and organized his information and data. He investigated the translations of the mathematical papyri, especially the Rhind Mathematical Papyrus, which was translated by Arnold Buffum Chace and housed at the British Museum. This was one of the three Hieratic Papyri that the author was able to examine, the other two being the Egyptian Mathematical Leather Roll and the Kahun Papyrus which are kept at the University College of London.

At the British Museum, the author also worked with three Demotic Papyri – the P. British Museum 10399, the P. British Museum 10520, and the P. British Museum 10794. He then traveled on to the Egyptian Museum in Cairo where he was given permission to photograph a complete set of the original Cairo Mathematical Papyrus (Appendix VII), one of the Demotic Papyri translated by Dr. Richard. A. Parker. The other Hieratic Papyri can be found in Moscow (the Mosco Papyrus) and in Berlin (the Berlin Papyrus and the Reisner Papyrus). The other Demotic Papyrus can be found in Copenhagen (the P. Carlsberg 30 Papyrus).

Using the Cairo Papyrus and its translation, this author was able to identify the drawings on the papyri for specific problems. Using his over 40 years of experience as a professor of mathematics, the author was able to determine the theorems and propositions used by the ancient mathematicians and written down by the scribes of the Egyptian papyri.

I am sure that when the Egyptologists and historians of mathematics have acquired a deeper understanding of the ancient Egyptians' contributions to mathematics, they will come to the rightful conclusion that the ancient Egyptians were farther advanced in the field of mathematics, in their time, than has been generally believed. Their contribution was essential in developing the field of mathematics.

In this newer edition we corrected all the mistakes that the scribes made not found earlier regarding the numerical errors and explained the mistakes in illustrations not matching the solutions. We will also show in greater detail the two methods for solving the problems used by the scribes – the mathematical rules and theorems, and the approximation method to find the estimated results. At this time they had a smaller store of mathematical facts and skills to solve the problems. Also, because of reader advice, we addressed certain points and issues that needed further clarification. Additionally, we added new information acquired since the previous printing. We will show the reader the mathematical concepts and theorems that presented themselves during the study of these papyri.

This book is a complete reference of mathematics that can be used by researchers, teachers, and students. It would also prove to be a valuable reference for college librarians and museums.

I am grateful to Mr. Sayed Hassan, the former Director of the Cairo Museum and the Curator of the museum's papyri. I am also thankful to the museum authorities for giving me permission to photograph and publish the photos of the Cairo Papyri.   My thanks also extend to the museum photographer and staff of the Division of the Papyri for all of their help and cooperation.

At this time, the author also would like to thank Ms. Deborah L. Mock, retired secretary for the Department of Communications Media at the Indiana University of Pennsylvania in the United States, who spent many, many hours, leading to years, editing and proofing this book.   I would also like to thank Ms. Delilah Black, a former student in the Center for Media Production and Research within the Communications Media Department at the Indiana University of Pennsylvania, who designed the cover for this book.   Last, but not least, I must thank my wife, Dr. Cheré Winnek-Shawer, who lovingly provided such excellent editorial advice and provided me with the encouragement and help that was needed these many years since the birth of this book.

Chapter 1
The Hieratic Mathematical Papyri

AHMES PAPYRUS, OR "THE RHIND MATHEMATICAL PAPYRUS" (RMP) (~1650 BC)

The Ahmes Papyrus was found during an illegal excavation at Thebes in the ruins of a small building near the Ramesseum, which is the memorial temple of Pharaoh Ramses II. It is better known as the Rhind Mathematical Papyrus (RMP) after the British collector, A. Henry Rhind, who purchased it in 1858. The Ramesseum is located in the Theban metropolis in Upper Egypt across the river from the modern city of Luxor. After Rhind's death, it came into the possession of the British Museum under BMP 10057 and BMP 10058.

The papyrus was written by a scribe named A'hmose, or Ahmes. The scribe says it is in the likeness of an older copy during the time of King "Ne-máet-Ré (Amen-em-hét III)," who reigned from 1849 BC to 1801 BC. The date when this copy was written is indicated by using the old Egyptian method of determining the year when a certain pharaoh reigned; in this case, "A-Wser-Ré," who was of the Hykos Dynasty, around 1650 BC. It was written in Hieratic and was a single roll nearly 18 feet long and about 13 inches high (Chace, 1927).

When the papyrus came to the British Museum, it was extremely brittle and some sections were missing. The missing parts were discovered and are now in the possession of the Brooklyn Museum in New York, USA. The missing parts were helpful in restoring it to its original form.

This papyrus is the most important source of the ancient Egyptian mathematics. It contains tables used to help in computations and 87 problems and their solutions on topics such as arithmetic, algebra, geometry, and trigonometry.

The tables are a division of two by fifty odd numbers from 3 to 101 and a table of division from 1-9 by 10, only in the unit fraction. It also contains hekat problems and problems related to volume, cylindrical granaries, division areas, circles, and circumscribing squares. There are also problems related to basic mathematics as used in recreation (mathematics and architecture). It also contains various methods of solutions, some of which have come down to the present day and are found in our arithmetic today.

The Rhind Papyrus was published in 1923 by Peet and contains discussion of the text that followed the translation and commentary by Griffith, Books I, II, and III. Chace published in 1927/29 with two volumes including photographs of the text. Robins and Shute published in 1987 and had a complete overview of the Rhind Papyrus.

In the Rhind Mathematical Papyrus, the 87 problems were divided by Chace into arithmetic, geometry, and miscellaneous problems (Chace, 1927). A complete list of this division is in Appendix 1. Thus we can say that the Rhind papyrus, while very useful to the Egyptian, was also "an example of the cultivation of mathematics as a pure science, even in its first beginnings." (Chace, 1927, pp. 42-43.)

THE MOSCO MATHEMATICAL PAPYRUS (MMP)

The Mosco Mathematical Papyrus is also called the Golenishchev Mathematical Papyrus after its first owner, Vladimir Semenoviâv Golenishchev, who acquired it in 1893. It later was entered into the collection at the Pushkin Museum of Art in Moscow as Inventory #4576.

The papyrus was written by an unknown scribe. It is approximately 18 feet long and varies between 1½ and 3 inches high. It probably dates back to the Thirteenth Dynasty and, based on older material, probably dated to the Twelfth Dynasty around 1850 BC (Clagett, 1999). It is older than the Rhind Mathematical Papyrus.

Its format was divided into 25 problems, with their solutions, by Vasilij Struve (1930). Six problems are unclear or unreadable (1, 2, 3, 11, 18, 23). Eleven problems are Pefsu problems (5, 8, 9, 12, 13, 15, 16, 20, 21, 22, 24). Three problems cover the area of the triangle, one of them (4) is to find the area of a right triangle and the other two (7, 17) are equivalent to the solutions of two simultaneous equations, one of the second degree. Two problems concern the solution of the equations of the first degree (19, 25). One problem has simultaneous equations, one of the equations of the second degree (6). See the details in Appendix II.

Two problems received interest among present-day mathematicians: Problem number 14 was to find the volume of a pyramid which is truncated such that the top area is square, its sides are 2 units in length, and the base area is square. Its side is 4 units and the verticle height is 6 units. The solution was that the volume is 56 units, which is correct. This shows that the ancient Egyptians knew the volume of the frustum. This is remarkable because the Greeks, at this point, did not solve this problem.

Also, Problem 10 involves computing the surface area of a Quonset hut type of roof, which is the earliest estimation of curvilinear area. The solution of this kind of problem needs modern calculus. These two problems will be discussed later in detail.

THE EGYPTIAN MATHEMATICAL LEATHER ROLL (EMLR) (~1650 BC)

The Egyptian Mathematical Leather Roll was discovered in some ruins of Ramesseum at Thebes in the same area the Rhind Papyrus was discovered. It was acquired by A. Henry Rhind, the same person who acquired the Rhind Papyrus.

The leather roll has been in the possession of the British Museum, BMP 10250, since 1864. The roll is 10 inches by 17 inches. It remained rolled for about 60 years when Dr. Alexander Scott and H. R. Hall unrolled it in 1927 and were able to restore the right hand column part which was damaged. It contains a table consisting of 26 decompositions of unit fractions. Scholars date the EMLR to the thirteenth century BC (Clagett, 1999). Glanville, (1927) regarded it as a "handy table for popular use." Glanville also said, "It must have been copied from a textbook as a table for future work" (Glanville, 1927, pp. 232-238). By studying this table and other tables of unit fractions, we can discover some of the general trends for the operation of addition of the unit fractions and the expansion of some unit fractions.

The 26 Egyptian fractions were converted to an Egyptian fraction series. The method for constructing this table will be discussed later, and the complete table and its mathematical analysis is in Appendix IV.

THE BERLIN MATHEMATICAL PAPYRUS (BMP)

The Berlin Papyrus was found at the ancient burial ground of Saqqara in the early 19th century AD. It was presented by Hans Schack-Schockenburg in two articles: Der Berlin Papyrus has Vol. 38 (1900) pp. 135-140 and Das Klurere Fragment de Berlin Papyrus 6619 (ibid Vol. 40 (1902)) pp. 65-66.

The papyrus was written in the Middle Kingdom according to C. Rossi (2007). According to Marshall Clagett (1999) it was written during the second half of the 12th or 13th Dynasty.

The papyrus contains ancient Egyptian mathematics and medical knowledge, including the first documentation concerning pregnancy testing procedures. It is kept in the Egyptian Museum in Berlin at Saatliche zie Berlin, Catalogue No. 6619.

The mathematics part contains two problems: given the area of the bigger square find, with some conditions, the size of the other two squares with their area equal to the area of the given square. This suggests some knowledge that was later known as the Pythagorean Theorem. It also shows a straight-forward solution of two second-degree equations, one unknown.

The simultaneous equations $x^2 + y^2 = a^2$ and $x = py$ are reduced to a single equation in y. These two problems will be explained later.

THE REISNER PAPYRI (RP)

The Reisner Papyri were found by Dr. George Reisner in 1904 at Nagéd Deir in Upper Egypt during an excavation carried out by Harvard University and the Boston Museum of Fine Arts, No. 38-2062, where it is now located. The four fragments of rolls were found lying in a wooden coffin. The first roll, RP1, was unrolled and restored by Academia der Wiessenchaften (Berlin), and the remaining rolls stayed in Berlin until the end of World War

II. These rolls, for the most part, represent the official register of a dockyard workshop. Dr. W. K. Simpson concluded it was written in the Twelfth Dynasty, about 1800 BC. Simpson divided the papyri into seventeen sections from A to Q. We are concerned with section I, subsections G, H, and I.

Section I, Subsection G: Subsection G consists of 19 lines of text. In the first, the column headings are given (length, width, depth, units, product/volume, and in the last column is the calculation of the number of workers needed for the work of that day.

Section I, Subsection H: Calculation of the stone blocks for a stone house. The format of the table is similar to subsection G, except the column heading product/volume is used and there is no column regarding the number of workers required.

Section I, Subsection I: Calculation of the floor plans, walls, trenches, and corridors. Subsection I records the length, width, height, and product/volume, with the unit presented as cubits except where the scribe mentions palms.

From this papyrus, we learn the relationship between the measurements they used: finger, palm, and cubit from a table which shows the relationship between these measurements.

It also contains a table dividing numbers 1 to 9 by 10, probably to help the builder construct the different chambers of the temple.

THE KAHUN MATHEMATICAL PAPYRUS (KMP)

The Kahun Mathematical Papyrus was found at Kahun (also known as El-Kahun) near the Pyramid of Sessostris by W. M. F. Petrie (grandson of the famous Australian explorer, Matthew Flinders). It is now kept at the University College of London, UC 32159 and UC 32160. This papyrus was judged to be written about 1750 BC. It contains six relatively small items of mathematical importance, some of which are translated and discussed by F. L. Griffith, University College of London. Also, some of it is explained by Schack-Schockenburg and R. J. Gillings.

A considerable amount of the papyrus is still not translated. One of the problems, (KPQV,3), presented in columns 11 and 12, will be explained by a complete analysis of the problem in Chapter 8. This example is where the distribution of the terms in the arithmetic progression appears to have been determined by a specific relationship between the larger term and the other terms in the progression.

Another problem is to compute the volume of a cylindrical granary. The scribe uses a formula with measurements in cubits and computes the volume, expressing it in terms of the unit khar, given the diameter (d) and height (h) of the cylindrical granary. These problems will be discussed later. There is also a table of Egyptian fractions shown in the form of $\frac{2}{n}$.

AKHMIN WOODEN TABLETS (AWT)

The Akhmin Wooden Tablets are two wooden tablets measuring 13 by 10 inches and covered in plaster. The tablets are inscribed on both sides. They date from an early Egyptian Kingdom and are now housed in the Cairo Museum in Egypt.

One side of the tablet has the name of the servant and the other contains mathematical texts. The first tablet contains three problems (division of the hekat by 13, 7, and 13), and the second tablet contains eleven problems – four of the division of the hekat by 11, 7, 3, 13, and 7 are on one side and division by 11, 11, 7, 11, 3, 7, and 10 on the other side.

The texts were reported by Daressy in 1901 and later analyzed and published in 1906.

The Akhmin Wooden Tablets posed a hekat unity $\left(\frac{64}{64}\right)$ five times. The unity was divided by 3, 7, 10, 11, and 13. The answer was written in a binary Horus Eye quotient and exact Egyptian remainder, scaled $\frac{1}{320}$ hekat, called ro.

In 2002, Hana Vymazaloya analyzed a fresh copy of the mathematical text and confirmed that the two-part answer of the fraction of the $\frac{1}{n}$ hekat is in terms of the Horus Eye fractions plus ro and its fractions.

The answer was checked several times by multiplying the fraction by 'n', and checking gave a result of 1 hekat.

A typographical error in Daressy's copy of two division of the hekat by 11 and 13 was corrected.

The problems will be analyzed later.

Chapter 2

The Hieratic Papyri Arithmetic

THE EGYPTIAN HIERATIC NUMERALS

1		10		100		1000	
2		20		200		2000	
3		30		300		3000	
4		40		400		4000	
5		50		500		5000	
6		60		600		6000	
7		70		700		7000	
8		80		800		8000	
9		90		900		9000	

There are different, but similar, variations of the writing of the Hieratic numerals, and the above table is one of these variations (Appendix VI).

Even though the papyrus is written in Hieratic, it is easier to explain in Hieroglyphics. The two systems use the same "base 10" and neither is positional. Most of the Hieratic Papyrus was translated to Hieroglyphics, and then from Hieroglyphics to English or French.

HIEROGLYPHIC NUMBER SYMBOLS

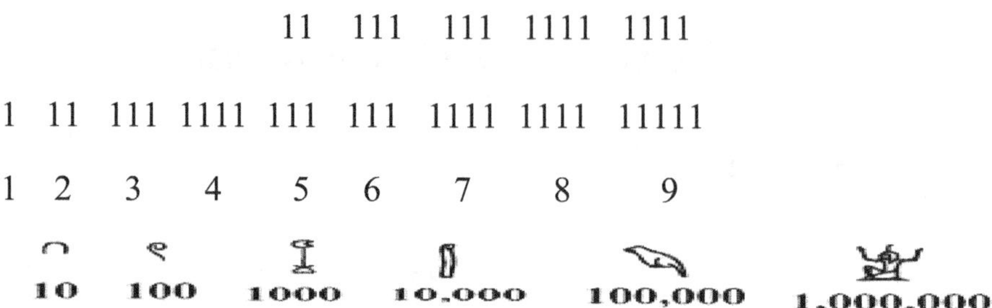

There were other symbols for larger units. The direction of writing was from right to left; for example, 2,576 was written as:

Since the position is not important in our writing, it is convenient to write the number from left to right instead of right to left, as is shown by taking the previous number and changing it to:

Sometimes small numbers require more characters than larger numbers; for example, 4,000 requires four characters, while 2,576 requires 20 characters. The system of writing uses base 10, but it is not positional.

The Hieratic number symbols allow the numbers to be written in a more compact form; for example, in Hieroglyphics, 9,999 was written in 36 symbols:

and in Hieratic, it was written in 4 symbols as shown:

However, from the previous table, it can be seen that the number in Hieratic required more symbols to be memorized.

Addition and Subtraction

The arithmetic was essentially additive. We can combine the similar symbols and replace every ten of them by the next higher order symbol. In Ahmes' Papyrus, the symbols for addition and for subtraction are:

addition: and subtraction: .

ADDITION

To add 2,467 to 2,634, the ancient Egyptians would do the following:

The result is:

Being as this is in Base 10:

11 ones is 1 ten leaving 1 one.

9 tens plus 1 ten equals 10 tens, which is 1 one hundred.

10 one hundreds plus 1 one hundred equals 1 one thousand, leaving 1 one hundred.

4 one thousands plus 1 one thousand becomes 5 one thousands.

This will equal 5,101.

SUBTRACTION

Subtraction is the reversal of the process of addition. If necessary, the Egyptians replaced the higher order by ten of the one order below it.

To subtract 325 from 1,433, the Egyptians would subtract: : ... : = (1)

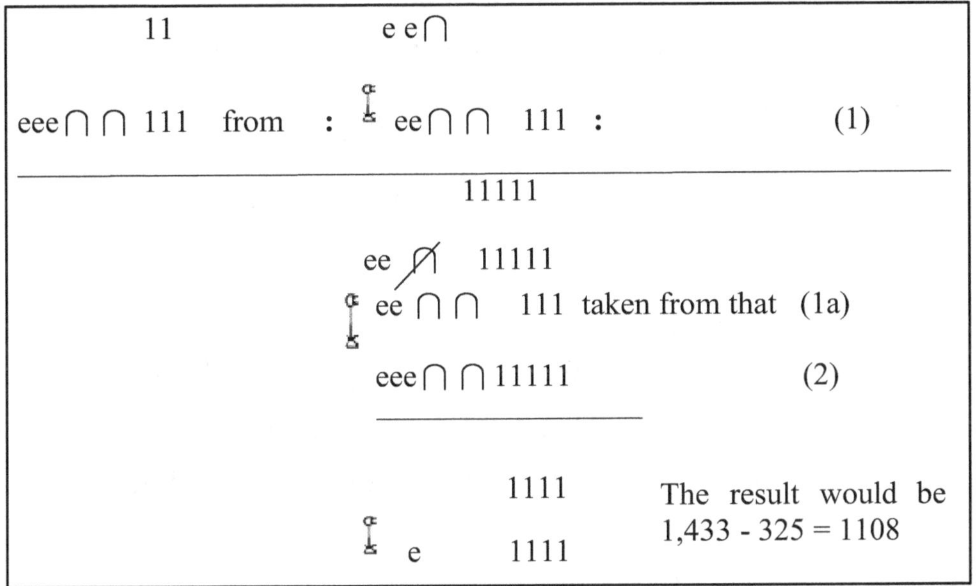

Because (1) has 3 ones and (2) has 5 ones, we can borrow 1 of the 10 from (1) and it will become (1a). Add 3 to it and it becomes 13. We took 5 from 13 and the result will be 8. (1111 1111).   (1a) has 2 tens and (2) has 2 tens; then the 10 place will be empty. (1a) has 4 hundreds and (2) has 3 hundreds. The result will be 1 hundred. (e). (1a) has 1 thousand and (2) has no thousands. The result will be 1 thousand. (   )

It is clear in hieroglyphics that addition and subtraction with smaller numbers may be performed by counting forwards and backwards. But with higher numbers, it is too awkward to count forward and backwards. Therefore, they may have had addition or subtraction tables like they had with multiplication and division tables (Appendix V).

MULTIPLICATION

According to the modern explanation of the known mathematical papyri, the foundation for the multiplication procedure was built on the series of 2^n, where n = 0, 1, 2, 3, 4, ... ; and the theorem that **any positive integer can be uniquely expressed as some of the sums of 2^n, where n = 0, 1, 2, 3, 4, ... ,** for some n and $2^0 = 1$ such as:

* $19 = 2^0 + 2^1 + 2^4$ $19 = 1+2+16$; $152 = 2^7 + 2^4 + 2^3$ $152 = 128 + 16 + 8$

When the Ancient Egyptian needed to multiply 51 by 19, he would first decide which is the multiplicand; assuming it is 51, then he would repeatedly multiply this by 2 (doubling), adding up some of the intermediate multipliers until they added up to the original multipliers, such as:

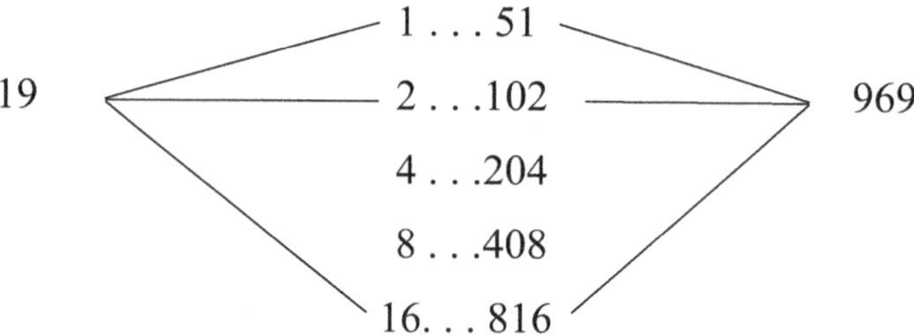

By adding numbers 1 + 2 + 16 = 19 and the corresponding numbers, 51 + 102 + 816 = 969, the results of the multiplication of 51 x 19 would be 969.

* $2^0 = 1$;   doubled is 2^1;   doubled is 2^2;   doubled is 2^3;   ...; and this process is called doubling.

DIVISION

The process of division was very close to the process of multiplication. If we want to divide 432 by 36, then by what number must we multiply 36 to get 432?

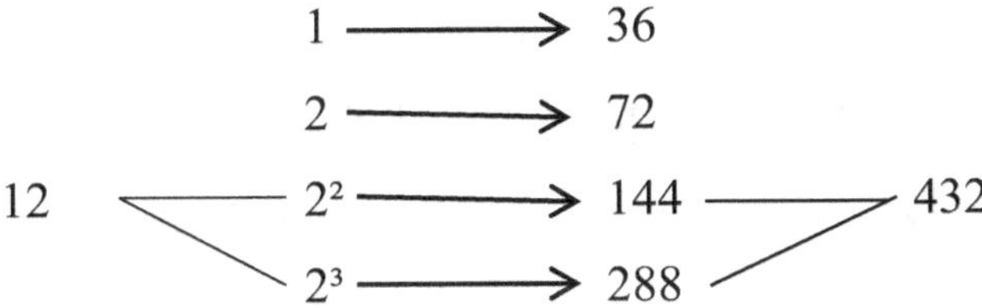

At this stage, we stop multiplying by 2 for the next doubling will be bigger than 432; however, if we add 144 to 288, we get 432. Therefore, 432 divided by $36 = 2^2 + 2^3 = 4 + 8 = 12$.

Chapter 3

Fractions

EGYPTIAN FRACTIONS, DEFINITION

An Egyptian fraction is the sum of a distinct unit fraction such as $\frac{1}{2}$, $\frac{1}{3}$, $\frac{1}{5}$, ..., and $\frac{2}{3}$, where each fraction denominator is different from the others.

The sum of an expression of this type is a positive fraction, shown as $\frac{1}{b}$.

Every positive fraction can be represented by the sum of some Egyptian fractions.

To write a unit fraction in Hieroglyphics, an Egyptian placed an 'open mouth', $\bigcirc$, over any number to represent the reciprocal of that number. For example:

$$\bigcirc_{|||} = \frac{1}{3} \qquad\qquad \bigcirc_{\cap\,||} = \frac{1}{12}$$

The Egyptian had special symbols for $\frac{1}{2}$, $\frac{2}{3}$, and $\frac{3}{4}$. These symbols are the following:

$$\underline{\diagdown} = \frac{1}{2}; \qquad\qquad \bigcirc_{\top} = \frac{2}{3}; \qquad\qquad \bigcirc_{\top} = \frac{3}{4}$$

HORUS EYE REPRESENTATION

Other representations of the fractions are the pieces of the Horus Eye. Legend has it that Horus went to war with his enemy Seth, but he had his eye gouged out and torn to pieces by his enemy. It was healed by Thoth. The ancient Egyptian used the pieces to represent fractions. The following diagram shows which part of the eye represented which fraction.

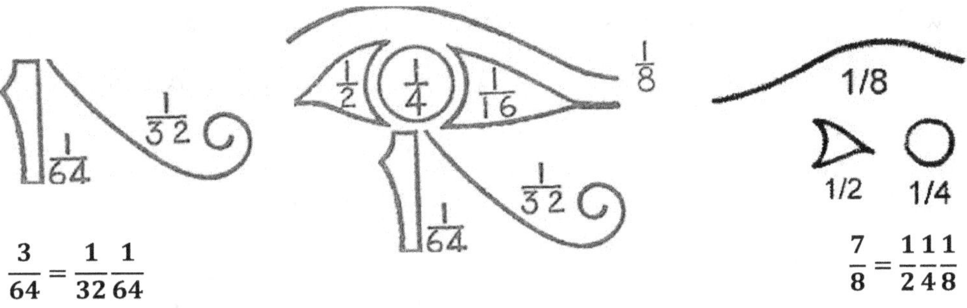

$$\frac{3}{64} = \frac{1}{32}\,\frac{1}{64}$$

$$\frac{7}{8} = \frac{1}{2}\,\frac{1}{4}\,\frac{1}{8}$$

In Hieratic, the open mouth became a dot. The dot was placed over the first digit of the number $i\wedge = \frac{1}{11}$. We will use the standard representation for the fraction, such as $\frac{a}{b}$ and the unit fraction is $\frac{1}{b}$.

The fraction $\frac{43}{48}$ was written as $\frac{1}{2}\,\frac{1}{3}\,\frac{1}{16} = \frac{1}{2} + \frac{1}{3} + \frac{1}{16}$. Each fraction has an infinite set of unit fractions. Consider the fraction $\frac{43}{48}$ and the identity

$$1 = \frac{1}{2} + \frac{1}{3} + \frac{1}{6}$$

We know that $\frac{43}{48} = \frac{1}{2} + \frac{1}{3} + \frac{1}{16}$.

Take the final fraction, $\frac{1}{16}$, and divide both sides of the identity by 16; then

$$\frac{1}{16} = \frac{1}{32} + \frac{1}{48} + \frac{1}{96}$$

then $\frac{43}{48} = \frac{1}{2} + \frac{1}{3} + \frac{1}{32} + \frac{1}{48} + \frac{1}{96}.$

We can repeat the process of expanding $\frac{1}{96}$ and so on.

Each time we get a different set of unit fractions which adds up to $\frac{43}{48}$ with fraction denominators different from each other. Some tables, such as an EMLR table, represent the fractions with more than one representation.

$$\frac{1}{8} = \frac{1}{10} + \frac{1}{40} = \frac{1}{12} + \frac{1}{24} = \frac{1}{15} + \frac{1}{25} + \frac{1}{75} + \frac{1}{200}.$$

Every positive rational number, $\frac{a}{b}$, can be represented by an Egyptian fraction. According to Boyer & Merzbach, 1989, **the sum of this type and similar sums, including $\frac{2}{3}$ and $\frac{3}{4}$ as summands, were used as series notations for rational numbers by the ancient Egyptians and continued to be used by other civilizations into medieval times.** Also, the Egyptian fractions were an object of study in modern number theory and recreational mathematics where they also had a practical application.

For example, we want to know which is larger, $\frac{5}{6}$ or $\frac{43}{48}$, without converting them into decimals. Now if each one was represented in Egyptian fraction form,

$$\text{then } \frac{5}{6} = \frac{1}{2} + \frac{1}{3} \text{ and } \frac{43}{48} = \frac{1}{2} + \frac{1}{3} + \frac{1}{16}.$$

Then it would be easy to know that $\frac{43}{48}$ is larger than $\frac{5}{6}$.

Another practical application: If we want to divide 5 loaves for 8 people, then we would put $\frac{5}{8}$ into Egyptian form. Then $\frac{5}{8} = \frac{1}{2} + \frac{1}{8}$ as shown in the picture:

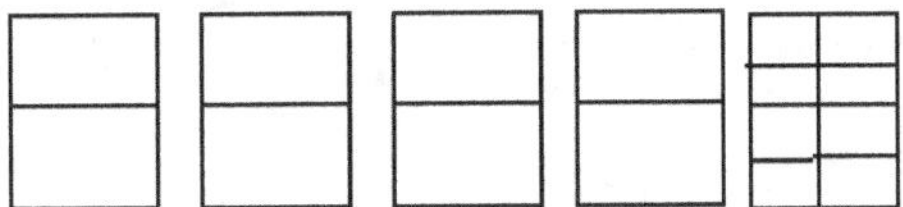

There are more people than loaves of bread so each person cannot get a whole loaf. But if we divide four loaves in half, we have one loaf left. Then they can divide that one loaf that is left into eight pieces and give each person $\frac{1}{8}$ loaf and the total is $\frac{5}{8} = \frac{1}{2} + \frac{1}{8}$, which is more practical than giving each person $\frac{5}{8}$ of a loaf of bread.

EXPANSION OF RATIONAL NUMBERS

If $\frac{a}{b}$ is a rational number and a < b, the fraction which is to be expanded is a unit fraction.

If a =1, then $\frac{a}{b} = \frac{1}{b}$

If a > 1 then it is a matter of dividing 'a' by 'b' and the result of the division is a unit fraction. What are the Egyptian fractions which, if multiplied by 'b', result in 'a'?

An example of the expansion of $\frac{6}{7}$ in Egyptian form:

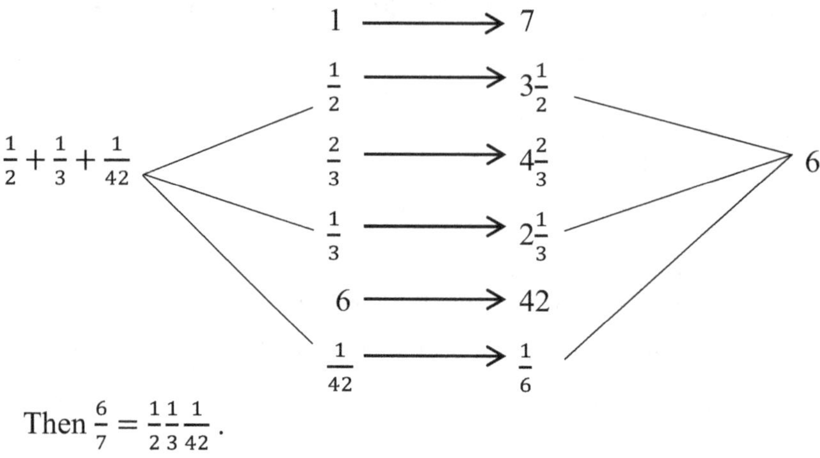

Then $\frac{6}{7} = \frac{1}{2}\,\frac{1}{3}\,\frac{1}{42}$.

The scribe used the rule that if a x b = c, then $\frac{1}{c}$ x b = $\frac{1}{a}$.

For example, if 6 x 7 = 42, then $\frac{1}{42}$ x 7 = $\frac{1}{6}$.

GREEDY ALGORITHM FOR COMPUTING EGYPTIAN FRACTIONS

Thus, every ordinary fraction has an Egyptian fraction form. In 1202 Fibonacci, in his book "Liber Abaci," suggested a greedy algorithm for computing Egyptian fractions. A greedy algorithm is an algorithm that follows the problem-solving heuristic of making the locally optimal choice at each stage with the hope of finding a global optimum, in which one repeatedly chooses the unit fraction with the smallest denominator that is not larger than the remaining fractions to be expanded. This means that in a modern notation, the fraction $\frac{x}{y}$ will be replaced by the following expansion:

$$\frac{x}{y} = \frac{1}{\left(\frac{y}{x}\right)} + \frac{-y \bmod x}{y\left(\frac{y}{x}\right)} = \frac{1}{\left(\frac{y}{x}\right)} + \frac{x\left(\frac{y}{x}\right) - y}{y\left(\frac{y}{x}\right)}$$

such that $\left(\frac{y}{x}\right)$ is the largest integer greater than $\frac{y}{x}$.

Example: $\dfrac{4}{13} = \dfrac{1}{\left(\frac{13}{4}\right)} + \dfrac{-13 \bmod 4}{13\left(\frac{13}{4}\right)} = \dfrac{1}{4} + \dfrac{3}{52}$

$\dfrac{3}{52} = \dfrac{1}{\frac{52}{3}} + \dfrac{-52 \bmod 3}{52\left(\frac{52}{3}\right)} = \dfrac{1}{18} + \dfrac{2}{52\times18} = \dfrac{1}{18} + \dfrac{1}{468}$

$\therefore \dfrac{4}{13} = \dfrac{1}{4} + \dfrac{1}{18} + \dfrac{1}{468}$.

This method always terminates with a finite expansion. To prove that,

$$\dfrac{x}{y} = \dfrac{1}{\left(\frac{y}{x}\right)} + \dfrac{x\left(\frac{y}{x}\right) - y}{y\left(\frac{y}{x}\right)}$$

$$\dfrac{x}{y} = \dfrac{1}{\left(\frac{y}{x}\right)} + \dfrac{x\left(\frac{y}{x}\right) - y}{x\left(\frac{y}{x}\right)} > 0$$

$$\dfrac{1}{\left(\frac{y}{x}\right)} < \dfrac{x}{y} \quad \text{then } x\left(\tfrac{y}{x}\right) - y > 0 \Rightarrow x\left(\tfrac{y}{x}\right) > y$$

$$\dfrac{1}{\left(\frac{y}{x}\right)-1} > \dfrac{x}{y} \quad \text{then } x\left(\tfrac{y}{x}\right) - x < y \Rightarrow x\left(\tfrac{y}{x}\right) - y < x .$$

Then $x\left(\tfrac{y}{x}\right) - y$ shows that the numerator of the remainder is smaller than the original numerator, and the denominator is getting larger. If the numerator of the remainder in its lowest term has a 1 in its top, we are finished; otherwise, we will repeat the process on the remainder, which has a smaller numerator, and see the remainder get smaller still when we take off its largest unit fraction. It will eventually stop at a numerator of 1 at some stage. The expansions are quite long with large denominators.

For example:

$$\dfrac{5}{121} = \dfrac{1}{25} + \dfrac{1}{757} + \dfrac{1}{763309} + \dfrac{1}{873960180913} + \dfrac{1}{1527612795642093418846225}$$

Fibonacci suggests another method of expansion, $\dfrac{a}{b}$, by replacing $\dfrac{a}{b}$ by $\dfrac{ac}{bc}$ such that:

$$\dfrac{b}{2} < c < b \text{ and expanding 'ac' as the sum of the divisor of 'ac'.}$$

To expand $\frac{5}{121}$, we can multiply the numerator and denominator by 33; then:

$$\frac{5}{121} = \frac{5 \times 33}{121 \times 33} = \frac{165}{121 \times 33} = \frac{121 + 44}{121 \times 33} = \frac{121}{121 \times 33} + \frac{44}{121 \times 33}$$

$$= \frac{1}{33} + \frac{4}{363} = \frac{1}{33} + \frac{3}{363} + \frac{1}{363}$$

$$= \frac{1}{33} + \frac{1}{121} + \frac{1}{363}$$

$$\therefore \frac{5}{121} = \frac{1}{33} + \frac{1}{121} + \frac{1}{363}$$

which is a much better expansion.

If 'n' can be written as the sum of divisors of 'N', then $\frac{n}{N}$ can be expanded to Egyptian fractions with no denominator greater than N itself.

For example: $\quad \frac{10}{21} = \frac{3+7}{21} = \frac{3}{21} + \frac{7}{21} = \frac{1}{7} + \frac{1}{3}$

EXPANDING FRACTIONS WITH PRACTICAL NUMBERS

Fibonacci, in his book "Liber Abaci," in connection with expanding the rational to an Egyptian fraction used a sequence of numbers, 1, 2, 4, 8, 12, 16, 18, 20, 24, ...which Srinivasan (1948) called practical numbers. This was the first attempted classification of these numbers that was completed by Stewart (1954) and Sierpiński (1955).

A practical number is a positive integer, 'N', such that for all n < N, n can be written as the sum of distinct divisors of N: For example, 20 is a practical number because its divisors are 1, 2, 4, 5, and 10, and all the numbers from 1 to 19 can be expressed as the sum of its divisors.

Therefore, we have $3 = 1 + 2$, $6 = 2 + 4$, $7 = 2 + 5$,

$8 = 1 + 2 + 5$, $9 = 4 + 5$, $11 = 1 + 10$, $12 = 2 + 10$, $13 = 1 + 2 + 10$,

$14 = 4 + 10$, $15 = 5 + 10$, $16 = 2 + 4 + 10$, $17 = 2 + 5 + 10$,

$18 = 10 + 5 + 2 + 1$, and $19 = 10 + 5 + 4$.

1. Every power of 2 is a practical number 2^n, such that 2, 4, 8, ...

2. Every even perfect number is a practical number that must have the form $2^{n-1}(2^n - 1)$.

3. Every primorial (the product of the first 'i' prime) where some 'i' is a practical number.

4. Any number that is a non-zero power of the first 'k' prime is a practical number. If $M < N$, then $d(M) < d(N)$. The number of the divisor of M is less than the number of the divisor of N. This includes Ramanugan's highly composite numbers, such as fractional numbers; an example being 1, 2, 4, 8, 12, 36, 48, 60, 120, 180, 240, 360, 720, 840, 1260, 1680, 2520, and 5040.

If the denominator is a practical number 'N' and $n < N$, then $\frac{n}{N}$ can be represented as an Egyptian fraction.

For example:

$$\frac{17}{24} = \frac{12+4+1}{24} = \frac{12}{24} + \frac{4}{24} + \frac{1}{24} = \frac{1}{2} + \frac{1}{6} + \frac{1}{24}$$

If 'n' is a practical number and 'N' is a relative prime with n and N < 2n, then 'qn' is also a practical number.

$\frac{9}{23}$, if choosing 12 as a practical number,

then N = 23 and n = 12, and 12, 23 is relatively prime

then 9 x 12 = 108   23 x 12 = 276

$$\frac{108}{276} = \frac{1}{3}\frac{1}{23}\frac{1}{69} .$$

Several methods of expansion involve algebraic identities for expanding the rational number to Egyptian fractions, such as:

1. $\frac{a}{ab-1} = \frac{1}{b} + \frac{1}{b(ab-1)}$

   For example:

   $$\frac{5}{14} = \frac{5}{15-1} = \frac{1}{3} + \frac{1}{3(14)} = \frac{1}{3} + \frac{1}{42}$$

2. $\frac{1}{n} = \frac{2}{3n} + \frac{2}{6n}$ and 'n' is even

   For example:

   $$\frac{1}{6} = \frac{2}{3 \times 6} + \frac{2}{6 \times 6} = \frac{1}{9} + \frac{1}{18}$$

More identities will be explained later.

From the tables for expanding the fractions found in the different papyri, the following are some of the observations noted in developing the tables, especially the $\frac{2}{n}$ table in the Rhind Mathematical Papyrus (RMP).

1) The smaller number of terms of the expansion is preferred and, in the $\frac{2}{n}$ table in the RMP, there never were more than 4 terms, such as

$$\frac{2}{101} = \frac{1}{101} + \frac{1}{202} + \frac{1}{303} + \frac{1}{606}$$

2) The unit fractions are always set down in descending order of the values of the fractions, such as $\frac{2}{15} = \frac{1}{10} + \frac{1}{30}$, and the same fraction is never used twice.

3) The smallest and first denominator is the main consideration, but the scribe accepts the larger first denominator if it will greatly reduce the last denominator.

4) Even numbers are preferred to odd numbers, even though they might be larger and even though the number of terms might be increased.

From this, the scribe was seeking the simplest values available in order to make them more convenient for calculation, and some mathematicians will call this kind of expansion an elegant expansion.

According to R. J. Gillings (1981), in 1967 Professor Hamblin, University of South Wales, used a computer program to calculate the unit fractions expressed in each division in the table of 2 by numbers 3, 5, ..., 101 in order to compare the decompositions given in the table with a thousand different possible forms. The time taken by the computer was 5 hours, and 22,295 values were produced by the computer.

For example, the computer found that there are 1,967 different decompositions of the limited fraction of $2 \div 45$ of a sum of not more than 4 terms, of which 7 have two terms, 134 have three terms, and 1,826 have four terms. The two terms are the following:

1. $\dfrac{2}{45} = \dfrac{1}{24} + \dfrac{1}{360}$

2. $\dfrac{2}{45} = \dfrac{1}{25} + \dfrac{1}{225}$

3. $\dfrac{2}{45} = \dfrac{1}{27} + \dfrac{1}{135}$

4. $\dfrac{2}{45} = \dfrac{1}{30} + \dfrac{1}{90}$

5. $\dfrac{2}{45} = \dfrac{1}{35} + \dfrac{1}{63}$

6. $\dfrac{2}{45} = \dfrac{1}{36} + \dfrac{1}{60}$

7. $\dfrac{2}{45} = \dfrac{1}{45} + \dfrac{1}{45}$

The scribe used $\dfrac{1}{30} + \dfrac{1}{90}$. Let us discuss why.

Number 7 was eliminated because the fraction was repeated. Numbers 2, 3, and 5 were also eliminated because they contained odd numbers. We can eliminate 1 because the second denominator is the largest of them all. Then we eliminate number 6 because the first fraction has a larger denominator than the first denominator in number 4, which was chosen by the scribe.

Many mathematicians have continued to study problems related to the Egyptian fraction including the expanding of special forms of fractions, the number of terms of expansion, etc. There are open problems such that Erodos-Straus conjectures concern that the length of the shortest expansion for fractions in the form: $\dfrac{4}{n} = \dfrac{1}{x} + \dfrac{1}{y} + \dfrac{1}{z}$ exists for every 'n'. It is known to be true for all $n < 10^{14}$, but the general truth of the conjecture is unknown.

RULES FOR CALCULATING THE $\frac{2}{n}$ TABLE (MODERN APPROACH)

In modern times, many attempts were made to discover the methods the Egyptians used in calculating the table. Most of these methods were described as algebraic identities; however, they do not match any single identity. Different methods were used for prime and for composite denominators.

Some of the following identities were suggested:

1) For small odd prime, (p) $\frac{2}{p} = \frac{2}{p+1} + \frac{2}{p(p+1)}$

$$= \frac{2}{7} = \frac{2}{8} + \frac{2}{7(8)} = \frac{1}{4} + \frac{1}{28}$$

The Hultsch-Bruins Method suggested that if 'p' is the large prime denominator, then $\frac{2}{p} = \frac{1}{a} + \frac{(2a-p)}{ap}$, where 'a' is a practical number. In the range $\frac{p}{2} < a < p$, the remaining number, 2(a-p), is a sum of the division of 'a' and forms a fraction $\frac{d}{ap}$ for each such division 'd' of the sum.

Example: $\frac{2}{41}$, if we choose 24 as a practical number, $\frac{41}{2} < 24 < 41$,

then $\dfrac{2 \times 24}{41 \times 24} = \dfrac{48}{41 \times 24} = \dfrac{41 + 7}{41 \times 24} = \dfrac{41}{41 \times 24} + \dfrac{7}{41 \times 24}$

$$= \frac{1}{24} + \frac{3+4}{41 \times 24}$$

$$= \frac{1}{24} + \frac{3}{41 \times 24} + \frac{4}{41 \times 24}$$

$$= \frac{1}{24} + \frac{1}{246} + \frac{1}{328}$$

For composite denominator factors, such as pq, then $\frac{2}{pq} = \frac{1}{aq} + \frac{1}{apq}$ when $a = \frac{p+1}{2}$.

For example:

$$\frac{2}{33} \qquad p = 3 \qquad q = 11 \qquad a = \frac{3+1}{2} = 2$$

$$\text{then } \frac{2}{33} = \frac{1}{2\times11} + \frac{1}{2\times3\times11} = \frac{1}{22} + \frac{1}{66}.$$

According to Gillings (1982) and Gardner (2002), these methods appear to have been used for many composite numbers in the Ahmes Papyrus, but there are exceptions such as $\frac{2}{35}$, $\frac{2}{91}$, and $\frac{2}{95}$.

$\frac{2}{pq}$ can be expanded as $\frac{1}{pr} + \frac{1}{qr}$, such that $r = \frac{p+q}{2}$. This will work for

$$\frac{2}{35} = \frac{2}{5\times7} \qquad r = \frac{5+7}{2} = 6$$

$$\frac{2}{35} = \frac{1}{5\times6} + \frac{1}{6\times7} = \frac{1}{30} + \frac{1}{42}.$$

Other methods use the property that $\frac{2}{pq} = \frac{1}{p}\left(\frac{2}{q}\right)$

for $\frac{2}{95} = \frac{1}{5}\left(\frac{2}{19}\right)$ from the table $\frac{2}{19} = \left(\frac{1}{12} + \frac{1}{76} + \frac{1}{114}\right)$

then $\frac{2}{95} = \frac{1}{5}\left(\frac{1}{12} + \frac{1}{76} + \frac{1}{114}\right) = \frac{1}{60} + \frac{1}{380} + \frac{1}{570}.$

The final prime expansion of the Ahmes table is

$$\frac{2}{101} = \frac{1}{101} + \frac{1}{101} \text{ since } 1 = \frac{1}{2} + \frac{1}{3} + \frac{1}{6}.$$

Then $\dfrac{1}{p} = \dfrac{1}{2p} + \dfrac{1}{3p} + \dfrac{1}{6p}$

$$\frac{1}{101} = \frac{1}{2(101)} + \frac{1}{3(101)} + \frac{1}{6(101)}$$

$$= \frac{1}{202} + \frac{1}{303} + \frac{1}{606}$$

then $\dfrac{2}{101} = \dfrac{1}{101} + \dfrac{1}{202} + \dfrac{1}{303} + \dfrac{1}{606}$.

RULES FOR FINDING TWO-THIRDS OF THE FRACTION

Problem 61 B (RMP)

This states that to get $\frac{2}{3}$ of $\frac{1}{5}$, take the reciprocals of 2 times 5 and

6 times 5, and in the same way get $\frac{2}{3}$ of the reciprocal of any odd number.

This means that if 'a' is odd, then $\dfrac{2}{3} \times \dfrac{1}{a} = \dfrac{1}{2a} + \dfrac{1}{6a}$ $\dfrac{2}{3} \times \dfrac{1}{5} = \dfrac{1}{10} + \dfrac{1}{30}$

Since $\dfrac{2}{3} = \dfrac{1}{2} + \dfrac{1}{6}$, then $\dfrac{2}{3}\left(\dfrac{1}{a}\right) = \dfrac{1}{2}\left(\dfrac{1}{a}\right) + \dfrac{1}{6}\left(\dfrac{1}{a}\right) = \dfrac{1}{2a} + \dfrac{1}{6a}$.

Also, the scribe used a similar rule that if $\frac{2}{3}$ of the reciprocal of any

even number is $\frac{1}{2a}$, then it will be the reciprocal of 3 times half the number,

such as $\dfrac{2}{3} \times \dfrac{1}{8} = \dfrac{1}{3 \times 4} = \dfrac{1}{12}$.

Problem 67 (RMP)

In the solution of the problem, the scribe found $\frac{2}{3}$ of $\frac{1}{3}$ as follows:

$$
\begin{array}{rcl}
1 & \longrightarrow & 1 \\[4pt]
\dfrac{2}{3} & \longrightarrow & \dfrac{2}{3} \\[6pt]
\dfrac{1}{3} & \longrightarrow & \dfrac{1}{3} \\[6pt]
\dfrac{2}{3}\ \text{of}\ \dfrac{1}{3} & \longrightarrow & \dfrac{1}{6}\ \ \dfrac{1}{18}
\end{array}
$$

The scribe applied the rules for $\frac{2}{3}$ of a fraction in the form $\frac{1}{a}$.

All Hieroglyphic and Hieratic fractions are unit fractions (unit numerators). The fraction $\frac{5}{6}$ is $\frac{1}{2}+\frac{1}{3}$ and fraction $\frac{23}{24}=\frac{1}{2}+\frac{1}{4}+\frac{1}{8}+\frac{1}{12}$. The exception to the unit numerators, as shown in problem 61A in the Rhind

Mathematical Papyrus (RMP), is the fraction $\frac{2}{3}$ which is denoted by ⊤ in

Hieroglyphic and ﻝ in Hieratic. They developed a prepared table for

finding $\frac{2}{3}$ of any fraction which has come to us from different papyri.

ADDITION OF FRACTIONS

By looking to the Egyptian Mathematical Leather Roll we found:

$$\frac{1}{9} + \frac{1}{18} = \frac{1}{6}$$

$$\frac{1}{12} + \frac{1}{24} = \frac{1}{8}$$

$$\frac{1}{24} + \frac{1}{48} = \frac{1}{16}$$

$$\frac{1}{21} + \frac{1}{42} = \frac{1}{14}$$

$$\frac{1}{15} + \frac{1}{30} = \frac{1}{10}.$$

It should be noted that the pattern or rule is **'if one unit fraction is double another and the largest denominator is divisible by 3, then the sum of the two fractions is a unit fraction with a denominator $\frac{1}{3}$ of the larger denominator.'**

This rule can be extended and generalized as follows:

'If one of the unit fractions is 'k' times the other, then their sum is the unit fraction, and its denominator is found by dividing the larger denominator by k+1 to provide the answer as an integer.'

The proof of the rule is the following:

Given any two fractions, $\frac{1}{a} + \frac{1}{b}$, if $b = ka$

$\frac{1}{a} + \frac{1}{ka} = \frac{(k+1)}{ka} = \frac{1}{n}$ where $n = \frac{b}{k+1}$ and assuming $\frac{b}{k+1}$ is an integer.

If $a = (k+1)m$

then $\frac{(k+1)}{k(k+1)m} = \frac{1}{mk}.$

Example: $\dfrac{1}{10}\dfrac{1}{40} = \dfrac{1}{8}$

$\dfrac{1}{10} + \dfrac{1}{40}$, k = 4; the result is $\dfrac{1}{8}$ where the 8 came from $40 \div 5 = 8$.

k + 1 = 5

We can use this rule to add more than two fractions, such as:

$\dfrac{1}{15} + \dfrac{1}{60} + \dfrac{1}{24} = \left(\dfrac{1}{15} + \dfrac{1}{60}\right) + \dfrac{1}{24}$.

For the first two, a = 60 = 4 × b   k = 4   a = 15 = (k+1)m = 5 × 3

m = 3 $\rightarrow \dfrac{1}{15} + \dfrac{1}{60} = \dfrac{1}{3\times4} = \dfrac{1}{12}$ Then $\dfrac{1}{12} + \dfrac{1}{24}$.

For $\dfrac{1}{12} + \dfrac{1}{24}$ then

a = 2 × 12 = 2b $\rightarrow$ k = 2   k + 1 = 3

b = 3 × 4 = (k + 1) 4 $\rightarrow \dfrac{1}{12} + \dfrac{1}{24} = \dfrac{1}{2\times4} = \dfrac{1}{8}$.

In general, the scribe used the following to add the fractions. If you wish to add two groups of fractions such as $\left(\dfrac{1}{3},\dfrac{1}{7}\right)$ and $\left(\dfrac{1}{5},\dfrac{1}{7},\dfrac{1}{21}\right)$, then the denominator has to be multiplied to produce the common denominator. We might think that we could add the two groups by writing $\left(\dfrac{1}{3},\dfrac{1}{5},\dfrac{1}{7},\dfrac{1}{7},\dfrac{1}{21}\right)$, but the law of the Egyptian fractions required that the fraction, $\dfrac{1}{7}$, could not be repeated. First, we find the common denominator which will be 105 and, if considering this as a number of loaves, then how many loaves is $\dfrac{1}{3}$ of 105 loaves? The answer will be 35 loaves and $\dfrac{1}{5}$ loaves of 105 loaves will be 21, $\dfrac{1}{7}$ loaves of 105 loaves will be 15, and $\dfrac{1}{21}$ loaves of 105 loaves will be 5. Then he adds 35+21+15+15+5 = 91 loaves. Then we want to find the part of 105 loaves which will make 91.

To summarize the previous:

$\frac{1}{3}$ of 105 is 35

$\frac{1}{7}$ of 105 is 15

$\frac{1}{5}$ of 105 is 21

$\frac{1}{7}$ of 105 is 15

$\frac{1}{21}$ of 105 is 5

____

Total: 91

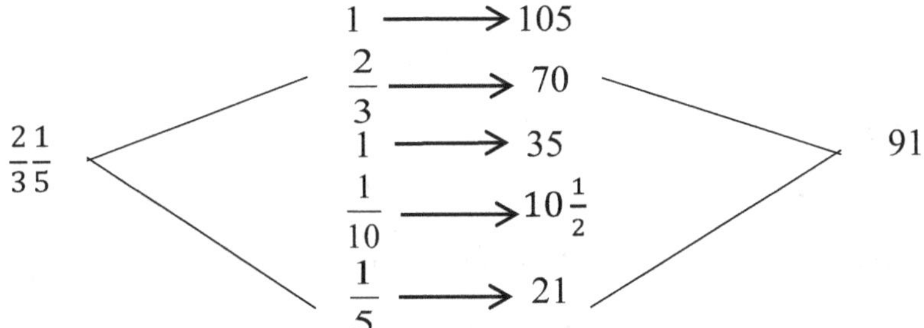

For him to get the 91 loaves, he needs to get $\frac{2}{3}$ of 105 = 70, and

$\frac{1}{5}$ of 105 = 21.

The answer for $\left(\frac{1}{3}, \frac{1}{5}, \frac{1}{7}, \frac{1}{7}, \frac{1}{21}\right) = \frac{2}{3}\frac{1}{5}$; but he might be able, without going to a formal multiplication, separate the 91 loaves into groups which one can recognize as part of the 105; thus, he might take 70 which is $\frac{2}{3}$ of 105 and 21 which is $\frac{1}{5}$ of the 105.

SUBTRACTION OF FRACTIONS

If they have an addition table, then subtraction can be found from the table.

If $\frac{1}{a} + \frac{1}{b} = \frac{1}{c}$, then $\frac{1}{c} - \frac{1}{a} = \frac{1}{b}$ and $\frac{1}{c} - \frac{1}{b} = \frac{1}{a}$

For example, if $\frac{1}{15} + \frac{1}{30} = \frac{1}{10}$, then $\frac{1}{10} - \frac{1}{15} = \frac{1}{30}$ and $\frac{1}{10} - \frac{1}{30} = \frac{1}{15}$;

or they might apply a rule such as:

If $a = kb$ and $b = (k-1)m$, then $\frac{1}{b} - \frac{1}{a} = \frac{1}{km}$.

For example:

$$\frac{1}{10} - \frac{1}{30} \qquad a = 3b \qquad (k-1)b = 2 \times 5 = 10$$

then $\quad k = 3 \qquad m = 5$

then $\dfrac{1}{10} - \dfrac{1}{30} = \dfrac{1}{3\times 5} = \dfrac{1}{15}$.

In general, the scribe used the following to subtract two fractions:

If we wish to subtract $\left(\frac{1}{5}, \frac{1}{7}, \frac{1}{21}\right)$ from $\left(\frac{1}{3}, \frac{1}{7}\right)$,

$\frac{1}{3}$ of 105 is 35
$\frac{1}{7}$ of 105 is 15 $\qquad$ 50
$\frac{1}{5}$ of 105 is 21
$\frac{1}{7}$ of 105 is 15 $\qquad$ 41
$\frac{1}{21}$ of 105 is 5

Take 41 from 50 the result is 9. $\quad 9 = 5+3+1$

$$\frac{9}{105} = \frac{5}{105} + \frac{3}{105} + \frac{1}{105} = \frac{1}{21} + \frac{1}{35} + \frac{1}{105}.$$

MULTIPLICATION OF FRACTIONS

Problem 8 (RMP)

Multiply $\frac{1}{4}$ by $1\frac{2}{3}\frac{1}{3}$

<u>The Solution</u>

$1 \rightarrow \frac{1}{4}$ as a part of 18, this is $4\frac{1}{2}$

$\frac{2}{3} \rightarrow \frac{1}{6}$ as a part of 18, this is 3

$\frac{1}{3} \rightarrow \frac{1}{12}$ as a part of 18, this is $1\frac{1}{2}$

Total: $\frac{1}{2}$ is 9 out of 18.

Problem 10 (RMP)

Multiply $\frac{1}{4}\frac{1}{28}$ by $1\frac{1}{2}\frac{1}{4}$

<u>The Solution</u>

$1 \rightarrow \frac{1}{4}\frac{1}{28}$ $\frac{1}{4}$ of 28 is 7 and $\frac{1}{28}$ of 28 is 1

$\frac{1}{2} \rightarrow \frac{1}{8}\frac{1}{56}$ $\frac{1}{8}$ of 28 is $3\frac{1}{2}$ and $\frac{1}{56}$ of 28 is $\frac{1}{2}$

$\frac{1}{4} \rightarrow \frac{1}{16}\frac{1}{112}$ $\frac{1}{16}$ of 28 is $1\frac{3}{4}$ and $\frac{1}{112}$ of 28 is $\frac{1}{4}$

Total $\rightarrow \frac{1}{2}$ is 14 out of 28

Then $(\frac{1}{4}\frac{1}{28})\,(1+\frac{1}{2}+\frac{1}{4}) = \frac{1}{2}$.

Note: In general, if the scribe wanted to add groups of fractions, he used methods to find a common denominator, but not necessarily the least common denominator (LCD). In some problems the scribe used the LCD, but scribes might often use a denominator that was smaller or larger than the LCD. If we used these methods, then some of the numerators of the fractions would not be integers

DIVISION OF FRACTIONS

Division is the reverse process of multiplication.

Divide $\frac{1}{2}$ by $\frac{1}{4}\frac{1}{28}$

The scribe will find a number which, if multiplied by $\frac{1}{4}\frac{1}{28}$, would result in $\frac{1}{2}$ as follows:

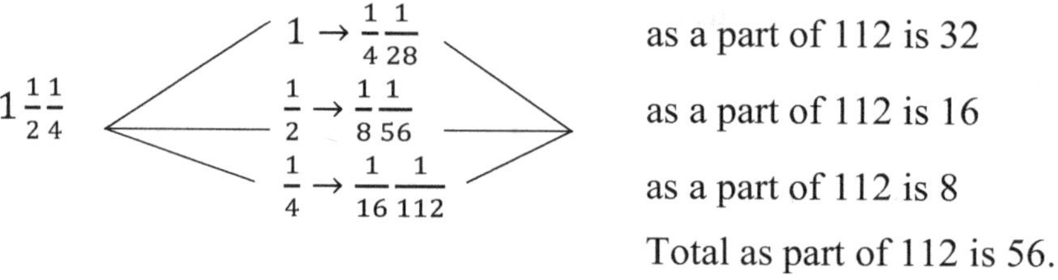

$1\frac{1}{2}\frac{1}{4}$:

$1 \rightarrow \frac{1}{4}\frac{1}{28}$ as a part of 112 is 32

$\frac{1}{2} \rightarrow \frac{1}{8}\frac{1}{56}$ as a part of 112 is 16

$\frac{1}{4} \rightarrow \frac{1}{16}\frac{1}{112}$ as a part of 112 is 8

Total as part of 112 is 56.

The answer is $1\frac{1}{2}\frac{1}{4}$.

Sometimes the scribe used the rule that if ab = c, then $\frac{1}{c}$(b) = $\frac{1}{a}$. In this case, any fraction can be used as an intermediate multiplier.

Example (from RMP):

Divide 2 by 53. The scribe's solution is as follows:

$$1 \longrightarrow 53$$
$$\frac{1}{30} \longrightarrow 1\frac{2}{3}\frac{1}{10}$$
$$6 \longrightarrow 318$$
$$\frac{1}{318} \longrightarrow \frac{1}{6} \longrightarrow 2$$
$$15 \longrightarrow 795$$
$$\frac{1}{795} \longrightarrow \frac{1}{15}$$

$$\frac{53}{30} = \frac{30}{30} + \frac{20}{30} + \frac{3}{30} = 1\frac{2}{3}\frac{1}{10}.$$

Remainder from 2 will be $\frac{7}{30} = \frac{5}{30} + \frac{2}{30} = \frac{1}{6} + \frac{1}{15}$.

To get $\frac{1}{6}$ and $\frac{1}{15}$, the scribe used the rule:

$ab = c$, then $\frac{1}{c}(b) = \frac{1}{a}$

$6 \times 53 = 318 \Rightarrow \frac{1}{318}(53) = \frac{1}{6}$

$15 \times 53 = 795 \Rightarrow \frac{1}{795}(53) = \frac{1}{15}$

The answer is $\frac{1}{30} + \frac{1}{318} + \frac{1}{795}$.

Ahmes may have used the previous procedure to expand $\frac{2}{n}$ for all odd positive integers $n = 1$ to $n = 101$ which was explained in detail in the Rhind Mathematical Papyrus (RMP). You can also find the entire table and its analysis in Appendix III.

Another method used, called scaling, is to find a number which multiplies the numerator and denominator by this number.

<u>Example:</u>

$\frac{2}{101}$ is scaled so that 2 x 6 = 12 and 101 by 6 = 606

12 out of 606 = 6 + 3 + 2 + 1 out of 606

$$6 \text{ out of } 606 = \frac{1}{101}$$

$$3 \text{ out of } 606 = \frac{1}{202}$$

$$2 \text{ out of } 606 = \frac{1}{303}$$

$$1 \text{ out of } 606 = \frac{1}{606}$$

$$\therefore \frac{2}{101} = \frac{1}{101} + \frac{1}{202} + \frac{1}{303} + \frac{1}{606}.$$

The Greek used the Egyptian fractions, which presided in Greece and Rome, well into the Medieval period. (Boyer & Merzbach, 1989, p. 24.)

Chapter 4

The Demotic Mathematical Papyri

Demotic refers to an ancient Egyptian script derived from the Hieratic script used in the Delta. The Demotic script was used by the Egyptians for more than a thousand years and, during that time, three stages of development occurred.

Early Demotic was used in lower Egypt during the 25th Dynasty. It dated between 600 and 400 BC. Most books written in Demotic are dated to the 26th Dynasty and the following Persian Period, the 27th Dynasty. After the unification of Egypt under Psametik, Demotic replaced Hieratic and became the official administrative and legal script for writing during the reign of Amasis. In the early period, the Demotic script resembled the Hieratic script.

Middle Demotic (400-30 BC) was the stage of writing used during the Ptolemaic Period. Books of literary texts and religious texts were written in Demotic. By the end of the 4th century BC, the Greek language became the official language in Egypt. Demotic lost its importance. In this period, the scripts had some similarity to the Greek scripts.

All the Demotic Mathematical Papyri were translated by Richard A. Parker in 1972, and his translation was used for the basis of studies of the Demotic Papyri.

THE CAIRO PAPYRUS E89127-30, 89137-43 (APPENDIX VII)

The papyrus was discovered at Tuna-El-Gabel (Hermopolis west) in 1938-1939. It is now located at the Cairo Museum, P. Cairo E89127-30, 89137-43. It is probably earlier than the third century BC, as claimed by Dr. Richard Parker, because the Demotic writing is close to the Hieratic writing. The papyrus is composed of eleven fragments, thus making the whole papyrus 2m long and 35cm in height. It also contains the Hermopolis legal code. Nineteen columns (A to S) have been established with certainty and are preserved in whole or in part. It is possible that columns K and L are missing.

Forty mathematical problems can be identified. Some are very fragmentary, and it is difficult to interpret. See the Cairo Papyrus photos in Appendix VII.

SUMMARY OF THE CAIRO PAPYRUS (CMP)

Problems 2 and 3 are to divide 100 by $11\frac{3}{4}$ and 100 by $15\frac{2}{3}$.

Problems 4 and 5 are composed of a series of fractions, and the scribe was asked to find the difference between the last fraction and the next to the last fraction.

In problems 7, 15, 16, and 18 we are given the area of a rectangle and the height is equal to 'k' times the width. These problems ask to determine the length and the width of the rectangle.

The scribe applies the same solution of the equation in problem LV,4 of the Kahun Papyrus which is:

if $xy = a$ and $x = ky$ then the solution he used is

$x = \sqrt{ak}$ and $y = \frac{x}{k}$.

In problems 8, 9, 10, 11, 12, 13, 14, and 17 the scribe was given the area and the height or the width of the piece of cloth. If one of the sides changes, what will be the other side if the area remains the same? If one side is 'a' and the other side is 'b', the theorem states that if $(a-h) = L$ and $(b+k)$ = M and LM = ab, then $M = b + \frac{hb}{a-h}$ and $L = a - \frac{ak}{b+k}$.

Problems 24, 25, 26, 27, 28, 29, 30, and 31 are pole problems. These poles stand against the wall, and the foot of the pole is moved outward a certain distance.

The scribe used the relationship between the lengths of the sides in the right-angle triangle that is formed, which we call the "Pythagorean Rule."

The scribe uses the formula: $a^2 = (a - x)^2 + y^2$, where 'y' is the distance that the pole was moved along the ground from the wall, and 'x' is the distance it is moved up or down on the wall, and 'a' is the length of the pole.

In problems 32 and 33, the scribe was given the area of the circle and was asked to determine the diameter.

The scribe introduced us to the relationship between the area of the circle (a), the diameter (d), and the circumference (c).

$d = \sqrt{a + \frac{a}{3}} \rightarrow d^2 = \frac{4}{3}a,$ if $d = 2r,$ r is the radius,

then $a = \frac{3}{4}(2r)^2 = 3r^2$ $\pi = 3$

Then the scribe introduces the circumference $c = 3d = 2\pi r$ and

$\frac{3}{4}d^2 = a = \frac{3d}{4} \times \frac{3d}{3} = \frac{c}{4} \times \frac{c}{3}.$

In problems 34 and 35, the scribe was given the area of the rectangle and the diagonal, and he was asked to determine the two sides.

The scribe used the solution of the following two equations where 'd' is the diameter and 'a' is the area.

$$x^2 + y^2 = d \quad \text{and} \quad xy = a$$

$$\text{which is} \quad x = \frac{1}{2}\left(\sqrt{d^2 + 2a} + \sqrt{d^2 - 2a}\right)$$

$$y = \frac{1}{2}\left(\sqrt{d^2 + 2a} - \sqrt{d^2 - 2a}\right), \text{to find x and y.}$$

In problem 36, the scribe was given a circular plot of land with an equilateral triangle inside with points touching the circle and a side of 'b'.

The scribe found the area of the triangle by finding the height, 'h', using the Pythagorean Theorem, then finding the area $a = \frac{1}{2} bh$. Then he found the height of the circular segment 'h*' equal to $\frac{1}{3}$ the height of the triangle, and he formed the area of one circular segment using the formula $a_2 = \left(\frac{h^* + b}{2}\right) h^*$. Next he found the area of the circular plot which equals $a_1 + 3a_2$. When proving his answer, he found the diameter by adding $\frac{1}{3}$ to the height of the triangle 'd', then he found c = 3d by using the formula $a = \frac{c}{3} \times \frac{c}{4}$.

Problem 37 is similar to problem 36 except the triangle was replaced by a square with a known side.

The area of the circular plot equals the area of the square + 4 (area of the segment).

Problem 38 is the same as 37 except that he was asked to determine the area of every part.

In problem 39, the height of a pyramid, 'h', and the side of the base, 'b', were given. The scribe was asked to determine the distance from the center of any side to the apex.

The scribe uses this formula to find the distance:

$$\text{The distance} = \sqrt{h^2 + \left(\frac{b}{2}\right)^2} \, .$$

Problem 40 was to find the volume, 'V', of the pyramid with a vertical, 'h', and the side of the base, 'b'.

The scribe uses the formula $V = \frac{1}{3} b^2 h$.

Any other problems which are not mentioned could not be identified from the papyrus.

THE ANALYSIS OF THE CAIRO PAPYRUS PROBLEMS

In problem 1, so little of the problem is preserved that we cannot identify its nature.

DIVISION PROBLEMS

In problems 2 and 3, the scribe applied the identity of

$$a \div b\frac{c}{d} = d\left(\frac{a}{bd+c}\right).$$

The proof for this identity is the following:

$$a \div b\frac{c}{d} = a \div \frac{bd+c}{d} = \frac{ad}{bd+c} = d\left(\frac{a}{bd+c}\right).$$

Problem 2

Divide 100 by $17\frac{2}{3}$ where a = 100; b = 17; c = 2; and d = 3.

The scribe multiplied $17\frac{2}{3}$ by a multiple of 3. He tried 6 and when he multiplied $6\left(17\frac{2}{3}\right)$ the result was 106. He estimated the answer to be less than 6.

To find the answer, he multiplied $3\left(17\frac{2}{3}\right)$ to find that $bd + c = 53$.

$100 \div 53 = \dfrac{a}{bd+c} = 1 + \dfrac{47}{53}$.

Then multiply $3\left(1 + \dfrac{47}{53}\right) = d\left(\dfrac{a}{bd+c}\right) = 5\dfrac{35}{53}$.

To check his answer, the scribe multiplied $17\frac{2}{3} \times 5\frac{35}{53}$.

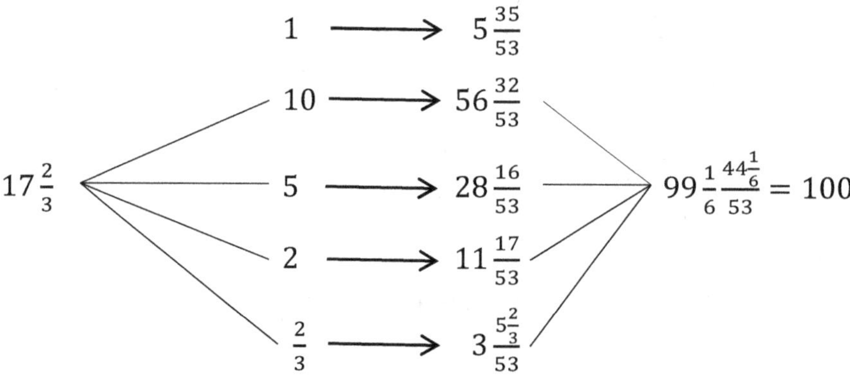

Problem 3

Divide 100 by $15\frac{2}{3}$ where a = 100; b =15; c = 2; and d = 3.

<u>The Solution</u>

The scribe took $6 \times 15\frac{2}{3}$; the result was 94. The answer is greater than 6.

To find the answer, the scribe took $3\left(15 + \frac{2}{3}\right)$ to find bd + c = 47.

$$100 \div 47 = \frac{a}{bd+c} = 2 + \frac{6}{47} = 2\frac{6}{47}.$$

To get these results, he took $3\left(2 + \frac{6}{47}\right) = d\left(\frac{a}{bd+c}\right) = 6 + \frac{18}{47} = 6\frac{18}{47}$.

<u>The Proof</u>

The scribe took $15\frac{2}{3} \times 6\frac{18}{47}$.

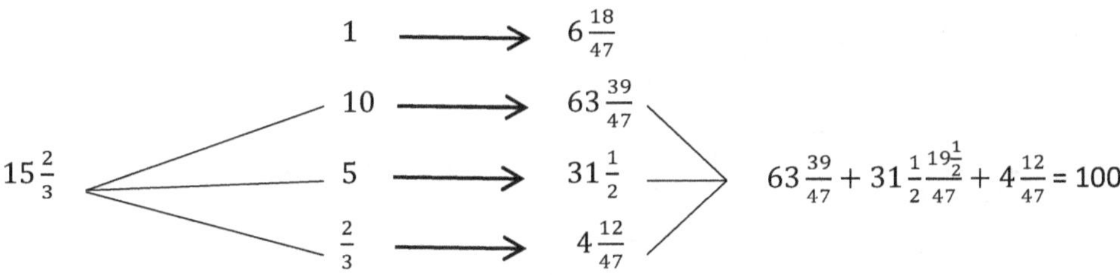

The scribe used the word 'prove' to mean 'checking the result.'

THE REMAINDER PROBLEMS

In **Problems 4 and 5**, the scribe applied the following:

If x = ab and y = cb, then $\frac{1}{x} - \frac{1}{y} = \frac{c-a}{a} \times \frac{1}{y}$.

To prove that, $\frac{1}{x} - \frac{1}{y} = \frac{1}{ab} - \frac{1}{cb} = \frac{c-a}{abc} = \left(\frac{c-a}{a}\right)\frac{1}{bc} = \left(\frac{c-a}{a}\right)\frac{1}{y}$.

Problem 4

Given a series $\frac{2}{3}, \frac{1}{6}, \frac{1}{60}, \frac{1}{120}$, and $\frac{1}{210}$ determine the remainder.

The scribe means the "remainder is the fraction resulting from subtracting the last fraction from the fraction before it."

The Solution

The scribe factored 120 and 210.

$x = 120 = 4 \times 30 = ab, \quad a = 4 \quad b = 30.$

$y = 210 = 30 \times 7 = bc, \quad c = 7 \quad b = 30.$

To get the result, he applied the formula, $\frac{1}{x} - \frac{1}{y} = \left(\frac{c-a}{a}\right)\frac{1}{y}.$

Then he found $c - a = 7 - 4 = 3.$

$\frac{c-a}{a} = \frac{3}{4}.$ Divide 4 by 3.

$$4 \quad\overset{\displaystyle 3 \longrightarrow 1}{\underset{\displaystyle 1 \longrightarrow \frac{1}{3} \longrightarrow 1\frac{1}{3}}{\diagdown}}$$

Multiply $210 \times 1\frac{1}{3}$; the result $= 280.$

The remainder is $\frac{1}{280} \quad \left(\frac{c-a}{a}\right)\left(\frac{1}{y}\right).$

<u>The Proof</u>

The scribe showed that $\frac{1}{280} + \frac{1}{210} = \frac{1}{120}$.

$$\frac{1}{280} \longrightarrow 1$$

$$\frac{1}{210} \longrightarrow 1\frac{1}{3}$$

Total: $2\frac{1}{3}$ to 1.

$$7 \longrightarrow 3$$
$$70 \longrightarrow 30$$
$$140 \longrightarrow 60$$
$$280 \longrightarrow 120$$

This shows that $280 = \frac{7}{3}(120)$ or $\frac{1}{280} + \frac{1}{210} = \frac{1}{120}$.

Note that in this problem, the scribe shows us that

$$\frac{1}{210} = \frac{4}{3}\left(\frac{1}{280}\right) \text{ and } \frac{1}{120} = \frac{7}{3}\left(\frac{1}{280}\right).$$

Problem 5

Given the series $\frac{5}{6}, \frac{1}{12}, \frac{1}{120}, \frac{1}{240}, \frac{1}{480}, \frac{1}{510}$, determine the remainder.

<u>The Solution</u>

Factor 480 and 510.

$$x = 480 = 16 \times 30 \qquad a = 16 \quad b = 30$$
$$y = 510 = 17 \times 30 \qquad c = 17 \quad b = 30$$

Apply the identity $\frac{1}{x} - \frac{1}{y} = \left(\frac{c-a}{a}\right)\left(\frac{1}{y}\right)$.

The scribe found $c - a = 1$; $\quad \frac{c-a}{a} = \frac{1}{16}$.

Then he multiplied 510×16; the result $= 8160$. The answer was $\frac{1}{8160}$.

<u>The Proof</u>

The scribe showed, as follows, that $\dfrac{1}{8160} + \dfrac{1}{510} = \dfrac{1}{480}$.

$$\dfrac{1}{8160} \longrightarrow 1$$

$$\dfrac{1}{510} \longrightarrow 16$$

$$\text{Total} \qquad 17$$

$$17 \longrightarrow 1$$
$$170 \longrightarrow 10$$
$$1700 \longrightarrow 100$$
$$3400 \longrightarrow 200$$
$$8160 \begin{cases} 6800 \longrightarrow 400 \\ 1360 \longrightarrow 80 \end{cases} 480$$
$$8160 \longrightarrow 480$$

Problem 6

This is not a problem, but an incomplete general statement of the procedure of adding to fractions with the largest denominator (smallest fraction) against the fraction of the smallest denominator (largest fraction), then putting the smallest fraction opposite to 1. (See the complete procedure above).

Problem 7

The scribe used the solution of 'x' and 'y' for the two simultaneous equations.

$$xy = a \quad (1)$$
$$\frac{x}{y} = b \quad (2) \quad \Longrightarrow \frac{y}{x} = \frac{1}{b};$$

which is $x = \sqrt{ab}$ and $y = \frac{x}{b}$.

<u>The Proof</u>

$$xy = a; \qquad \frac{x}{y} = b \Longrightarrow y = \frac{x}{b} = \left(\frac{1}{b}\right)x \; ;$$

$$\therefore x\frac{x}{b} = a \Longrightarrow x^2 = ab \Longrightarrow x = \sqrt{ab} \quad \text{and} \quad y = \frac{\sqrt{ab}}{b}.$$

The scribe also used $\sqrt{a^2 + b} \approx a + \frac{b}{2a}$ or $\sqrt{a^2 - b} = a - \frac{b}{2a}$.

Problem 7

Given 1000 cloth cubits (square cubits) of rectangular sails with the height $1\frac{1}{2}$ times the width, determine the height and the width.

Let x = height and y = width;

a = the area of the rectangle = xy = 1000 cloth cubits;

$b = 1\frac{1}{2}; \; \frac{1}{b} = \frac{2}{3}$.

<u>The Solution</u>

Take $ab = 1000\left(1 + \frac{1}{2}\right) = 1500$ cloth cubits.

Find $x = \sqrt{ab} = \sqrt{1500} \approx 38\frac{2}{3}\frac{1}{20} \approx$ height in cubits

Then $= \frac{2}{3}x = \frac{2}{3}\left(38\frac{2}{3}\frac{1}{20}\right) \approx 25\frac{2}{3}\frac{1}{10}\frac{1}{90} \approx$ width in cubits.

The Proof

Not given by the scribe, but

$$\left(38\,\tfrac{2}{3}\tfrac{1}{20}\right)\left(25\,\tfrac{1}{10}\tfrac{1}{90}\right) = 998\,\tfrac{4}{135} \approx 1000 \text{ in square cubits.}$$

Note: $\sqrt{1500} \approx \sqrt{1444+56} = \sqrt{(38)^2+56} = 38 + \tfrac{56}{78} \approx 38\,\tfrac{2}{3}\tfrac{1}{14}\,+,$

or $\sqrt{1500} \approx \sqrt{1521-21} = \sqrt{(39)^2-21} = 39 - \tfrac{21}{76} \approx 38\,\tfrac{2}{3}\tfrac{1}{16}\,-,$

The scribe chose $\sqrt{1500} \approx 38\,\tfrac{2}{3}\tfrac{1}{20}$.

In **Problems 8-12**, the scribe used the solution for the equation

h = height;   w = width;   a = area; and x = the value subtracted from 'h'; and

'y' = the value added to 'w.'

If $hw = a$ (1); $(h-x)(w+y) = a$ (2); and $(h-x)w = a_1$(3);   then $y = \dfrac{a-a_1}{h-x}$.

The Proof

$hw = a;$ $(h-x)w = a_1$

$a - a_1 = hw - (h-x)w$

$\qquad = hw - hw + xw = xw$

Let $hw = (h-x)(w+y) = a$.

Then $hw = hw + hy - xw - xy;$ $xw = y(h-x);$ $\Rightarrow y = \dfrac{xw}{h-x} = \dfrac{a-a_1}{h-x}.$

Problem 8

Given a piece of cloth 7 cubits high by 5 cubits wide, determine what must be added to the width when 1 cubit is taken from the height and the area stays the same.

<u>The Solution</u>

h = height = 7; w = width = 5; $x = 1$;   and a = area

$a = 7 \times 5 = 35$ cloth cubits; $h - x = 7 - 1 = 6$;

$a_1 = (h - x)w = 6 \times 5 = 30.$

The answer: $y = \dfrac{a - a_1}{h - x} = \dfrac{5}{6}$ cubit.

$$w + y = 5\frac{5}{6} \text{ cubits.}$$

<u>The Proof</u>

$$a = 6\left(5\frac{5}{6}\right) = 35 \text{ cloth cubits.}$$

Problem 9

Given a piece of cloth 6 cubits high by 4 cubits wide, what must be added to the width when 1 cubit is taken from the height and the area remains the same?

<u>The Solution</u>

$a = h \times w = 6 \times 4 = 24$ cloth cubits; $x = 1$.

The new height $= 6 - 1 = 5 = h - x$ (we took $\frac{1}{6}$ of the height).

$a_1 = 5 \times 4 = 20$; $a - a_1 = 24 - 20 = 4$.

The answer, $\frac{a-a_1}{h-x} = \frac{4}{5} = y$, the value to be added to the width in cubits.

Note the scribe used $\frac{2}{3}\,\frac{1}{10}\,\frac{1}{30}$ instead of $\frac{4}{5}$. The new width is $4\frac{2}{3}\,\frac{1}{10}\,\frac{1}{30}$ cubits.

<u>The Proof</u>

$$a = 5\left(4\frac{2}{3}\,\frac{1}{10}\,\frac{1}{30}\right) = 24 \text{ cloth cubits.}$$

Problem 10

Given a piece of cloth 6 cubits high and $1\frac{1}{2}$ cubits wide, determine what must be added to the width when $\frac{1}{2}$ cubit is taken from the height and the area remains the same.

<u>The Solution</u>

The scribe used the same procedure from Problem 9.

$a = h \times w = 6 \times 1\frac{1}{2} = 9$ cloth cubits.

Subtract $\frac{1}{2}$ from the height.

The result is $5\frac{1}{2}$ cubits.

To make it easy, he multiplied the length by 2 to make 12 (h^*), and the width by 2 to make $3 = w^*$.

$$h^* - x^* = 12 - \frac{1}{12}h^* = 11; \qquad a^* = 12 \times 3 = 36 .$$

$$a_1^* = 11 \times 3 = 33; \quad \text{then} \quad a^* - a_1^* = 36 - 33 = 3 .$$

The answer is $y^* = \frac{a^* - a_1^*}{h^* - x^*} = \frac{3}{11}$ and the result $= 3\frac{3}{11} = w^* + y^*$.

Take $\frac{1}{2}$ of $3\frac{3}{11} = 1\frac{1}{2}\frac{1\frac{1}{2}}{11} = y$.   The new width $= 1\frac{1}{2}\frac{1\frac{1}{2}}{11}$ cubits.

The Proof

$$a = 5\frac{1}{2}\left(1\frac{1}{2}\frac{1\frac{1}{2}}{11}\right) = 9 \text{ cloth cubits.}$$

In **Problems 11, 12, and 14**, the scribe applied the solution of the equations:

$$hw = a \quad (1); \quad h(w - y) = a_1 \quad (2); \quad \text{and} \quad (h + x)(w - y) = a \quad (3);$$

then $x = \frac{a - a_1}{w - y}$.

The Proof

$$a - a_1 = hw - h(w - y) = hw - hw + hy = hy$$

since $hw = a$ and $(h + x)(w - y) = a$

$$hw = hw - hy + xw - xy$$

$$\Rightarrow hy = x(w - y) \Rightarrow x = \frac{hy}{w - y} = \frac{a - a_1}{w - y} .$$

The new height $= h + \frac{hy}{w - y} = \frac{hw - hy + hy}{w - y} = \frac{hw}{w - y} = \left(\frac{w}{w - y}\right)h = h \div \left(\frac{w - y}{w}\right).$

Problem 11

Given a piece of cloth 6 cubits high and 4 cubits wide, determine what must be added to the height when 1 cubit is taken from the width and the area remains the same.

<u>The Solution</u>

The scribe solved the problem by using the result of the solution of the three equations for 'x'.

a = area; h = height = 6 cubits; w = width = 4 cubits;

x = the increase of h; y = the decrease of w

a = area = $h \times w = 6 \times 4 = 24$ cloth cubits.

a_1 = new area = $h(w - y) = 6 \times 3 = 18 \Rightarrow a - a_1 = 24 - 18 = 6$ and $w - y = 4 - 1 = 3$ cubits.

$a = (h + x)(w - y);$ then $x = \dfrac{a - a_1}{w - y} = \dfrac{24 - 18}{3} = 2$ cubits.

The new height is $h = 6 + 2 = 8$ cubits.

<u>The Proof</u>

$a = 8 \times 3 = 24$ cloth cubits.

Problem 12

Given a piece of cloth 6 cubits high and 3 cubits wide, determine the height when $1\frac{1}{2}$ cubits are taken off the width and the area remains the same.

The Solution

The scribe solved the problem using the result of the solution of the three equations for 'x' as follows:

a = area; h = height; w = width;

x = the increase of h; and y = the decrease of w .

$a = 6 \times 3 = 18$ cloth cubits;

$$w - y = 3 - 1\frac{1}{2} = 1\frac{1}{2} \text{ and } \frac{w-y}{w} = \frac{1\frac{1}{2}}{3} = \frac{1}{2}$$

The new height is $6 \div \frac{1}{2} = 6 \times 2 = 12$ cubits $\Rightarrow h \div \left(\frac{w-y}{w}\right) = \frac{hw}{w-y}$.

The Proof

$a = 12 \times 1\frac{1}{2} = 18$ cloth cubits.

Problem 13

Most of the problem was lost.

Problem 14

Given a piece of cloth 21 cubits high and 5 cubits wide, determine what must be added to the width when 1 cubit is taken from the height, and the area remains the same.

The Solution

By using the previous general solution of the three equations for 'y', it follows that:

$$h = 21; \quad w = 5; \quad x = 1 \qquad 21 \times 5 = 105 \quad hw = a;$$

$$21 - 1 = 20 \quad h - x.$$

$$\frac{5}{20} = \frac{1}{4} \qquad\qquad \frac{xw}{h-x}$$

The new width is $5 + \dfrac{1}{4} = 5\dfrac{1}{4} = w + \dfrac{xw}{h-x}$ cubits.

The Proof

$$a = 20\left(5\tfrac{1}{4}\right) = 105 \quad \text{cloth cubits.}$$

For **Problems 15, 16, 17, and 18** the scribe applied the solution of the two simultaneous equations which he applied in Problem 7.

$$\text{If } a = hw \quad (1) \qquad \text{and} \quad h = pw \quad (2)$$

$$\text{then } h = \sqrt{pa} \qquad \text{and} \quad w = \frac{1}{p}\sqrt{pa}$$

(see **Problem 7** on page 49).

Problem 15

Given a piece of cloth with an area of 1000 square cubits with a height $1\frac{1}{2}$ times the width, determine the height and the width.

The Solution

$a = 1000$ cloth cubits $\qquad p = 1\frac{1}{2} \qquad\qquad 1000\left(1\frac{1}{2}\right) = 1500\ = ap$

$\sqrt{1500} = 38\,\frac{2}{3}\,\frac{1}{20}$ cubits. The height is $\sqrt{ap}$.

$\frac{2}{3}\left(38\,\frac{2}{3}\,\frac{1}{20}\right) = 25\,\frac{2}{3}\,\frac{1}{10}\,\frac{1}{90}$ cubits. The width is $\frac{1}{p}\sqrt{ap}$.

The Proof

Not found by the scribe.

If we multiply $\left(38\,\frac{2}{3}\,\frac{1}{20}\right) \times \left(25\,\frac{2}{3}\,\frac{1}{10}\,\frac{1}{90}\right)$, the result is $998\,\frac{1}{45}\,\frac{1}{135}$ square cubits, with a difference of about .197% resulting from taking the square root of 1500.

Problem 16

Given a piece of cloth with an area of 100 square cubits, and the height is in a ratio of 7 to 5 of the width, determine the height and width of the cloth.

The Solution

The scribe used the previous procedure as follows:

$$a = 100 \text{ cloth cubits} \qquad p = \frac{7}{5} = 1 + \frac{2}{5}$$

$$\frac{2}{5} = \frac{1}{3}\frac{1}{15} \qquad 100\left(\frac{1}{3}\frac{1}{15}\right) = 40$$

$$100 \times \frac{7}{5} = 100\left(1 + \frac{2}{5}\right) = 100 + 100\left(\frac{1}{3}\frac{1}{15}\right) = 140 = ap$$

$$\sqrt{140} = 11\frac{5}{6} = 11\frac{2}{3}\frac{1}{30}\frac{1}{70} \qquad \sqrt{ap} = \text{height in cubits}$$

$$\frac{5}{7}\left(11\frac{2}{3}\frac{1}{30}\frac{1}{120}\right) = 8\frac{1}{3}\frac{1}{10}\frac{1}{60} \qquad \frac{1}{p}\sqrt{ap} = \text{width in cubits}$$

The Proof

$$a = \left(11\frac{2}{3}\frac{1}{30}\frac{1}{120}\right)\left(8\frac{1}{3}\frac{1}{10}\frac{1}{60}\right) = 99\frac{5}{6}\frac{1}{10}\frac{1}{20}\frac{1}{120} \text{ cloth cubits.}$$

The remainder (error) is $\frac{1}{20}$.

Problem 17

Given a piece of cloth 21 cubits high by 5 cubits wide, determine the height when 1 cubit is taken from the width and the area remains the same.

The Solution

The scribe solved this problem by following the general pattern he used in Problems 11 and 12.

a = area; h = height; w = width;
x = the increase of h; and y = the decrease of w .
$h = 21$ $w = 5$ $y = 1$ $x = 7$
$a = 21 \times 5 = 105$ cloth cubits $= hw$ $5 - 1 = 4 \; = \; w - y$
$21 \div 4 \; = \; 5\frac{1}{4}$ $\dfrac{hy}{w-y} = x$ cubits.
The new height is $21 + 5\frac{1}{4} = 26\frac{1}{4}$ cubits.

The Proof

$$a = \left(26\frac{1}{4}\right)4 = 105 \quad \left(h + \frac{hy}{w-y}\right)(w - y) = hw \;\; \text{cloth cubits.}$$

Problem 18

Given a piece of cloth with an area of 100 square cubits with a height that has a ratio of 20 to 2 of the width, determine the height and the width of the cloth.

The Solution

The scribe used the same pattern he used in Problem 7.

$$a = 100 \text{ cloth cubits} \qquad p = \text{height to the width} = \frac{20}{2} = \frac{10}{1} = 10$$

$$100 \times 10 = 1000 \; = ap$$

$$\text{height} = \sqrt{1000} = 31\frac{1}{2}\frac{1}{10}\frac{1}{30} \text{ cubits;} \qquad\qquad \sqrt{ap} \; .$$

$$\text{width} = \frac{1}{10}\sqrt{1000} \approx 3\frac{1}{6} = \left(3\frac{1}{10}\frac{1}{20}\frac{1}{100}\frac{1}{300}\right) \text{ cubits;} \qquad \frac{1}{p}\sqrt{ap}.$$

<u>The Proof</u>

The scribe did the following:

$$\left(31\tfrac{1}{2}\tfrac{1}{10}\tfrac{1}{30}\right)\left(3\tfrac{1}{6}\right) = 100\tfrac{1}{6}\tfrac{1}{180} \text{ cloth cubits.}$$

The error is $\tfrac{1}{6}\tfrac{1}{180}$, resulting from the approximation.

Note: The scribe used $\sqrt{1000} = \sqrt{1024 - 24}$ rather than $\sqrt{961 + 36}$, and the width was rounded from $3\tfrac{49}{300}$ to $3\tfrac{1}{6}$ cubits.

Problem 19, 20, and 21

These problems were not clear according to the translator. The problems concerned calculations involving interest. We cannot complete or reconstruct them because too much is missing.

Problem 22 is completely missing.

CONTINUOUS HALVING OF GIVEN NUMBERS

Problem 23

This problem is a mathematical exercise involving continuous halving of given numbers. The following is a summary of the scribe's writing.

The first: 1

The second: $\tfrac{1}{2}$ $\qquad\qquad 1 + \tfrac{1}{2} + \tfrac{1}{4} + \tfrac{1}{8} = 1\tfrac{7}{8} = \tfrac{15}{8}$

The third: $\tfrac{1}{4}$

The fourth: $\tfrac{1}{8}$

The total: $1\tfrac{5}{6}\tfrac{1}{30}\tfrac{1}{120} \qquad 1 + \tfrac{5}{6} + \tfrac{1}{30} + \tfrac{1}{120} = 1\tfrac{7}{8} = \tfrac{15}{8}$

To get $\tfrac{1}{8}$ of the total, we multiply by $\tfrac{1}{15}$.

The result is $\tfrac{1}{15} + \tfrac{1}{18} + \tfrac{1}{450} + \tfrac{1}{1800} = \tfrac{1}{8}$ or $\tfrac{1}{15} + \tfrac{1}{30} + \tfrac{1}{60} + \tfrac{1}{120}$.

The scribe tried to show us different methods to expand $\frac{1}{8}$.

Half of $\frac{1}{15}$ is $\frac{1}{30}$.

Take $8 \times \frac{1}{30} = \frac{6}{30} + \frac{2}{30} = \frac{1}{5}\frac{1}{15}$.

First: $\qquad \frac{1}{5}\frac{1}{15}$

Second: $\qquad \frac{1}{10}\frac{1}{30}$

Third: $\qquad \frac{1}{15}$

Fourth: $\qquad \frac{1}{30}$

The scribe tried to show us that any fraction multiplied by 8 and then divided by 2 three times resulted in the original fraction.

Note: **Problems 24 to 31**, the pole problems, the scribes used the Pythagorean Theory. It should be noted an exact parallel to these problems is to be found on a Babylonian tablet from the Kassite period or earlier (thus before 1200 BC). (R. Parker, 1972, p. 6) Pythagoras came to Egypt during the time of King Amasis to further his studies. King Amasis introduced him first to the Priest of Heliopolis, then to the Priest of Memphis, and lastly to the Priests of Thebes, to each of whom Pythagoras gave a silver goblet. He was finally initiated with their secrets. His attainments in geometry corresponded with the ascertained fact that Egypt was the birth place of that science (Philarch de Repugn. Stoic 2 p. 1089; Demetrius; Antisthenes; Cicero de Natura Decorum III, 36P.) (James, 1954, p. 43.)

POLE PROBLEMS (USE OF THE PYTHAGOREAN RULE)

Problem 24, 25, and 26

These problems are concerned with an erect pole with a foot that moves outward a certain distance. The problem is to determine how far the top of the pole was lowered if the bottom of the pole was moved outward. The scribe used the Pythagorean Theorem as follows:

a = the hypotenuse (the length of the pole)

a = the height of the wall before the pole's foot was moved outward.

$$a^2 = (a - x)^2 + y^2$$

x = the decrease in height

y = the amount the foot is moved out.

$$a^2 = (a - x)^2 + y^2$$

$$(a - x)^2 = a^2 - y^2$$

$$a - x = \sqrt{a^2 - y^2}$$

$$x = a - \sqrt{a^2 - y^2}\,.$$

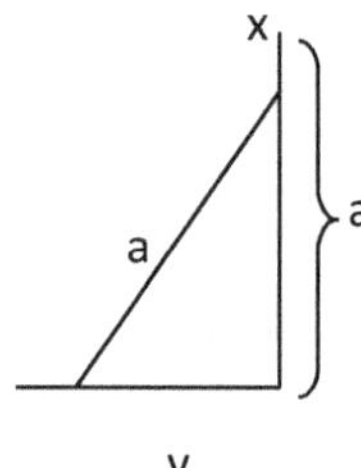

Problem 24

A pole is 10 cubits when erect and its foot is moved outward 6 cubits. By what amount is the top of the pole lowered?

$a = 10;$ $y = 6;$ $x =$ the number of cubits the top is lowered.

The Solution

$a =$ length of the pole; $y =$ distance the foot is moved outward

$a^2 = 10 \times 10 = 100$

$y^2 = 6 \times 6 = 36$

$100 - 36 = 64$ $a^2 - y^2 = (a - x)^2$

$\sqrt{64} = 8$ $a - x = \sqrt{(a^2 - y^2)}$

$10 - 8 = 2$ $x = a - \sqrt{(a^2 - y^2)}$

$x =$ amount top of the pole is lowered $= 2$ cubits.

Problem 25

The erect pole is $14\frac{1}{2}$ cubits high, and its foot is moved 10 cubits outwards. What is the amount that the top of the pole was lowered?

$a = 14\frac{1}{2};$ $y = 10;$ $x =?$

The Solution

$14\frac{1}{2} \times 14\frac{1}{2} = 210\frac{1}{4} = a^2$

$10 \times 10 = 100 = y^2$

$210\frac{1}{4} - 100 = 110\frac{1}{4} = a^2 - y^2 = (a - x)^2$

$\sqrt{110\frac{1}{4}} = 10\frac{1}{2} = \sqrt{a^2 - y^2} = a - x$

$14\frac{1}{2} - 10\frac{1}{2} = 4;$ $x = a - \sqrt{a^2 - y^2}$

The top of the pole is lowered by 4 cubits.

Problem 26

The erect pole is 10 cubits high and its foot is moved outwards 8 cubits. What is the amount by which the pole is lowered?

$$a = 10; \quad y = 8; \quad x = a - \sqrt{a^2 - y^2}$$

The Solution

$10 \times 10 = 100 = a^2$

$8 \times 8 = 64 = y^2$

$100 - 64 = 36 = a^2 - y^2$

Reducing 36 to its square root is $6 = \sqrt{a^2 - y^2}$

$10 - 6 = 4 \qquad\qquad x = a - \sqrt{a^2 - y^2}$

The top of the pole was lowered by 4 cubits.

Problems 27, 28, and 29 had given information as follows:

The length of the pole $= a$;　the number of cubits the pole was moved downwards $= x$. We are to find the distance that the foot of the pole was moved outwards (y).

The Solution

The scribe used the Pythagorean Theorem to find 'y'.

$$y = \sqrt{a^2 - (a - x)^2}$$

The Proof

$a^2 = (a - x)^2 + y^2$

$y^2 = a^2 - (a - x)^2$

$y = \sqrt{a^2 - (a - x)^2}$

Problem 27

The erect pole is 10 cubits high. How far is the foot of the pole moved outward when the top of it is moved downward by 2 cubits?

$a = 10;$ $x = 2;$ $y =?$

The Solution

$10 \times 10 = 100 = a^2$
$10 - 2 = 8 = a - x$
$8 \times 8 = 64 = (a - x)^2$
$100 - 64 = 36 = a^2 - (a - x)^2$

Reducing 36 to its square root $= 6 \ = \sqrt{a^2 - (a - x)^2} = y$

The foot of the pole was moved outward 6 cubits.

Problem 28

The erect pole is $14\frac{1}{2}$ cubits high. How far is the foot of the pole moved outward if the top of it is moved 4 cubits downward?

$a = 14\frac{1}{2};$ $x = 4;$ $y =?$

The Solution

$14\frac{1}{2} \times 14\frac{1}{2} = 210\frac{1}{4} = a^2$

$14\frac{1}{2} - 4 = 10\frac{1}{2} = a - x$

$10\frac{1}{2} \times 10\frac{1}{2} = 110\frac{1}{4} = (a - x)^2$

$210\frac{1}{4} - 110\frac{1}{4} = 100 = a^2 - (a - x)^2$

100, reduced to its square root, is $10 \ = \sqrt{a^2 - (a - x)^2} = y$

The base of the pole was moved outward 10 cubits.

Problem 29

The erect pole is 10 cubits high. How far was the foot of the pole moved outward if the top of the pole was moved downward 4 cubits?

$$a = 10; \qquad x = 4; \qquad y = ?$$

The Solution

$10 \times 10 = 100 = a^2$

$10 - 4 = 6 = a - x$

$6 \times 6 = 36 = (a - x)^2$

$100 - 36 = 64 = a^2 - (a - x)^2$

Reducing 64 to its square root $= 8 = \sqrt{a^2 - (a - x)^2} = y$

The foot of the pole was moved outward 8 cubits.

In **Problems 30 and 31**, the scribe applied the solution for the equation $x^2 = (x - k)^2 + L^2$ which is $x = \dfrac{k^2 + L^2}{2k}$.

The Proof

According to the Pythagorean Theorem, the proof is:

$x^2 = (x - k)^2 + L^2$

$x^2 = x^2 - 2 \times k + k^2 + L^2$

$2 \times k = k^2 + L^2$

$x = \dfrac{k^2 + L^2}{2k}$.

In these two problems, we are given the number of cubits by which the pole is lowered 'k' and the number of cubits by which the foot of the pole was moved outward 'L'. Find the length of the pole 'x'.

Problem 30

Given that the top of an erect pole has been lowered 2 cubits, and the foot of the pole has been moved outward 6 cubits, determine the original height of the pole.

The Solution

$$6 \times 6 = 36 = L^2$$
$$2 \times 2 = 4 = k^2$$
$$36 + 4 = 40 = L^2 + k^2$$
$$2 \times 2 = 4 = 2k$$
$$40 \div 4 = 10 = \frac{L^2 + k^2}{2k} = x$$

The original height of the pole is $x = 10$ cubits.

Problem 31

Given that the top of an erect pole has been lowered 4 cubits, and the foot of the pole has been moved outward 10 cubits, determine the original height of the pole.

The Solution

$$10 \times 10 = 100 = L^2$$
$$4 \times 4 = 16 = k^2$$
$$100 + 16 = 116 = L^2 + k^2$$
$$2 \times 4 = 8 = 2k$$
$$116 \div 8 = 14\frac{1}{2} = \frac{100+16}{8} = \frac{L^2+k^2}{2k} = x$$
$$x = 14\frac{1}{2}$$

The original height of the pole was $14\frac{1}{2}$ cubits.

THE AREA OF A CIRCLE

Rules for the Relationship of Diameter (d), Circumference (c), and Area (a).

In the Hieratic Papyri, the formula for the area of the circle is

$$a = \left(\frac{8}{9}d\right)^2$$

The diameter is $d = 2r$ where r is the radius of the circle.

$$a = \pi r^2$$

$$\pi r^2 = \left(\frac{8}{9}d\right)^2 = \left(\frac{8}{9}\right)^2 (2r)^2$$

$\pi \approx 3.16049$. Compare that to 3.141592654.
The error $\approx .6\%$.

In the Demotic Papyri, the scribe assumed the following:

$$d = \sqrt{a + \frac{a}{3}} \qquad a = \text{area of circle; diameter, } d = 2r \quad (1)$$

$$d^2 = \frac{4}{3}a \Longrightarrow a = \frac{3}{4}d^2 \Longrightarrow a = \frac{3}{4}(4r^2) = 3r^2 \qquad (2)$$

Comparing the area of the circle,

$$\pi r^2 = 3r^2 \therefore \pi = 3.$$

The error is 4.5%.

The circumference is $c = 3d$ $\qquad\qquad\qquad$ (3)

$$\text{and } = \frac{3}{4} \times \frac{3}{3}d^2 = \frac{(3d)(3d)}{4\times 3} = \frac{c}{4} \times \frac{c}{3} \qquad\qquad (4)$$

$$\text{and } a = \frac{3}{4}d^2 = d^2 - \frac{d^2}{4}.$$

In the Demotic Papyri, the scribe used the rules 1, 2, 3, and 4 in dealing with the circle.

Problem 32

Given a circular piece of land with an area of 100 square cubits, determine the diameter.

The Solution

The area of the circle is 100 square cubits. This will be 'a'.

$100 + \frac{1}{3}$ of that area is $133\frac{1}{3}$ square cubits, which is 'a $+ \frac{1}{3}$a'.

Reducing to its square root d $= 11\frac{1}{2}\frac{1}{20}$ cubits. $d = \sqrt{a + \frac{1}{3}a}$.

The Proof

$3\left(11\frac{1}{2}\frac{1}{20}\right) = \left(34\frac{1}{2}\frac{1}{10}\frac{1}{20}\right) =$ the circumference in cubits. $\qquad \frac{c}{3}$

$\frac{1}{3}\left(34\frac{1}{2}\frac{1}{10}\frac{1}{20}\right) = 11\frac{1}{2}\frac{1}{20}$ $\qquad \frac{c}{3}$

$\frac{1}{4}\left(34\frac{1}{2}\frac{1}{10}\frac{1}{20}\right) = 8\frac{1}{2}\frac{1}{10}\frac{1}{20}\frac{1}{120}$ $\qquad \frac{c}{4}$

$\left(11\frac{1}{2}\frac{1}{20}\right)\left(8\frac{1}{2}\frac{1}{10}\frac{1}{20}\frac{1}{120}\right) = 100$ square cubits. $\qquad \frac{c}{3}\frac{c}{4} = a$

Note: $\frac{1}{120}$ is rounded off to the correct number of $\frac{1}{80}$.

The correct answer is $100\frac{3}{800}$ square cubits.

The error is $\frac{1}{400}\frac{1}{800}$ square cubits.

Problem 33

Given a circular piece of land (converted from a square) with an area of 10 square cubits, determine the diameter.

The Solution

$$10 + \frac{1}{3}10 = 13\frac{1}{3} = a + \frac{1}{3}a$$

The square root of $13\frac{1}{3} = 3\frac{2}{3}$ $\qquad\qquad d = \sqrt{a + \frac{1}{3}a}$

$3\frac{2}{3}$ is the diameter of the piece of land.

Its circumference is $\left[3 \times \left(3\frac{2}{3}\right)\right] = 11$ cubits $\qquad c = 3d$

The Proof

$\frac{1}{3} \times 11 = 3\frac{2}{3}$ $\qquad\qquad\qquad \frac{c}{3}$

$\frac{1}{4} \times 11 = 2\frac{1}{2}\frac{1}{4}$ $\qquad\qquad\qquad \frac{c}{4}$

$a = 3\frac{2}{3} \times 2\frac{1}{2}\frac{1}{4} = 10$ square cubits $\qquad a = \frac{c}{3} \times \frac{c}{4}$

The approximate result is $10\frac{1}{12}$ square cubits.

In **Problems 34 and 35**, the scribe applied the solution for 'x' and 'y' in the following equations:

$$x^2 + y^2 = d^2 \quad \text{and} \quad xy = a$$

$$x = \frac{1}{2}\left(\sqrt{d^2 + 2a} + \sqrt{d^2 - 2a}\right) \quad \text{and} \quad y = \frac{1}{2}\left(\sqrt{d^2 + 2a} - \sqrt{d^2 - 2a}\right)$$

<u>The Proof</u>

$$x^2 + y^2 = d^2$$

$$xy = a \longrightarrow 2xy = 2a$$

$$x^2 + y^2 + 2xy = d^2 + 2a$$

$$(x + y)^2 = d^2 + 2a$$

$$x + y = \sqrt{d^2 + 2a} \qquad (1)$$

$$x^2 + y^2 - 2y = d^2 - 2a$$

$$(x - y)^2 = d^2 - 2a$$

$$x - y = \sqrt{d^2 - 2a} \qquad (2)$$

Add (1) and (2)...

$$2x = \sqrt{d^2 + 2a} + \sqrt{d^2 - 2a}$$

$$\therefore x = \frac{1}{2}\left(\sqrt{d^2 + 2a} + \sqrt{d^2 - 2a}\right)$$

Subtract (2) from (1)...

$$2y = \sqrt{d^2 + 2a} - \sqrt{d^2 - 2a}$$

$$\therefore y = \frac{1}{2}\left(\sqrt{d^2 + 2a} - \sqrt{d^2 - 2a}\right).$$

Area of a Rectangle

Problem 34

Given a rectangle with an area (xy) = 60 square cubits and a diagonal of d = 13, determine the two sides with height = x and width = y

$$a = xy = 60 \qquad d = 13 \qquad d^2 = x^2 + y^2 = 169$$

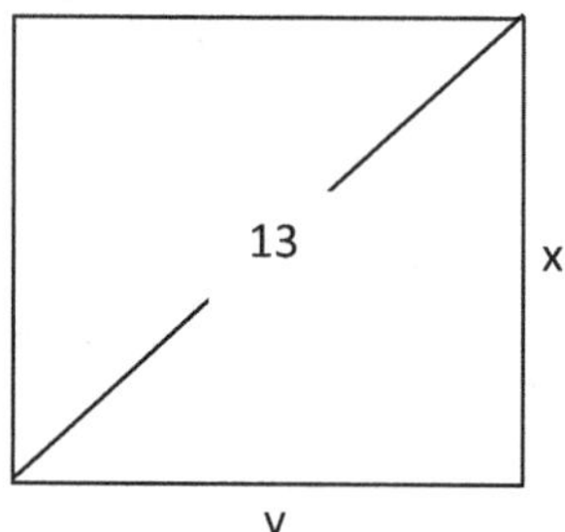

<u>The Solution</u>

$$13 \times 13 = 169 = d^2$$

$$60 \times 2 = 120 = 2a$$

$$169 + 120 = 289 = d^2 + 2a.$$

Reduce 289 to its square root and the result is $17 = \sqrt{d^2 + 2a}$
$$169 - 120 = 49 = d^2 - 2a.$$

Reduce 49 to its square root and the result is $7 = \sqrt{d^2 - 2a}$

$$2x = 17 + 7 = 24 = \sqrt{d^2 + 2a} + \sqrt{d^2 - 2a}$$

$$x = \frac{1}{2} \text{ of } 24 = 12 = \text{height} = \frac{1}{2}\left(\sqrt{d^2 + 2a} + \sqrt{d^2 - 2a}\right) \text{ cubits.}$$

$$2y = 17 - 7 = 10 = \sqrt{d^2 + 2a} - \sqrt{d^2 - 2a}$$

$$y = \frac{1}{2} \text{ of } 10 = 5 = \text{width} = \frac{1}{2}\left(\sqrt{d^2 + 2a} - \sqrt{d^2 - 2a}\right) \text{ cubits.}$$

The answer is that the land is 12 cubits by 5 cubits in size.

<u>The Proof</u>

$$x^2 = 12 \times 12 = 144$$

$$y^2 = 5 \times 5 = 25$$

$$x^2 + y^2 = 144 + 25 = 169$$

Reduce 169 to its square root of 13 cubits.

$$d = 13 \text{ cubits.}$$

Problem 35

Given a rectangle with an area of 60 square cubits and a diagonal of 15 cubits, determine the two sides.

$$a = xy = 60 \qquad d = 15 \qquad x^2 + y^2 = 225$$

$$x = \tfrac{1}{2}\left(\sqrt{d^2 + 2a} + \sqrt{d^2 - 2a}\right) \quad \text{and} \quad y = \tfrac{1}{2}\left(\sqrt{d^2 + 2a} - \sqrt{d^2 - 2a}\right)$$

<u>The Solution</u>

Using the same procedure as in Problem 34,

$$15 \times 15 = 225 = d^2$$

$$60 \times 2 = 120 = 2a$$

$$225 + 120 = 345 = d^2 + 2a$$

Reduce 345 to its square root of $18\tfrac{1}{2}\tfrac{1}{12}$ $\qquad\qquad \sqrt{d^2 + 2a}$

$$225 - 120 = 105 = d^2 - 2a$$

Reduce 105 to its square root and get $10\tfrac{1}{4}$ $\qquad\qquad \sqrt{d^2 - 2a}$

The width $= \tfrac{1}{2}\left(\sqrt{d^2 + 2a} - \sqrt{d^2 - 2a}\right)$ cubits.

$$= \tfrac{1}{2} \text{ of } 8\tfrac{1}{3} = 4\tfrac{1}{6} \text{ cubits.}$$

The height $= \tfrac{1}{2}\left(\sqrt{d^2 + 2a} + \sqrt{d^2 - 2a}\right)$ cubits.

$$= \tfrac{1}{2}\left(18\tfrac{1}{2}\tfrac{1}{12} + 10\tfrac{1}{4}\right) = 14\tfrac{1}{3}\tfrac{1}{12} \text{ cubits.}$$

The Proof

$$14\frac{1}{3}\frac{1}{12} \times 14\frac{1}{3}\frac{1}{12} = 207\frac{5}{6}$$

$$4\frac{1}{6} \times 4\frac{1}{6} = 17\frac{1}{6}$$

$$207\frac{5}{6} + 17\frac{1}{6} = 225$$

Reduce 225 to its square root. The result is 15 cubits which is the diagonal.

AREA OF CIRCLE

(Using a Triangle or a Square within the Circle and the Segment of the Circle)

Problem 36

Given a circular plot of land in which is laid out an equilateral triangle with

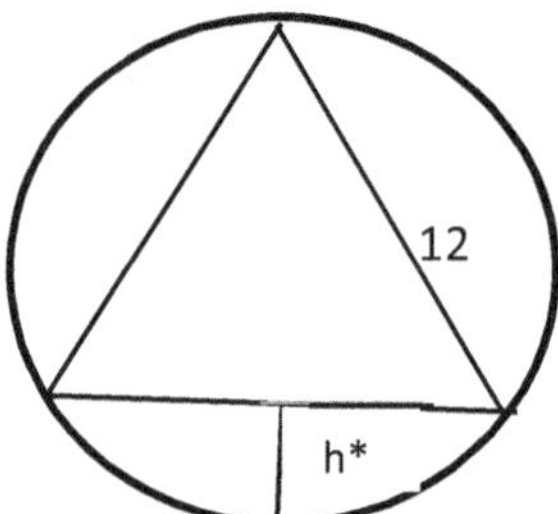

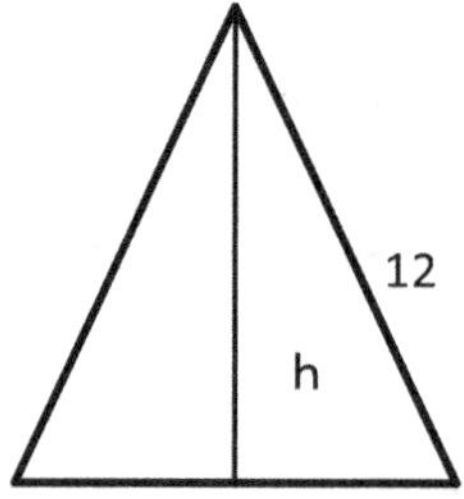

sides 12 divine cubits each, determine the area of the plot.

The scribe found the height of the triangle and multiplied it by half of the base $\left(a = \frac{1}{2}bh\right)$; then he found the area of the segment using the formula $a = \left(\frac{h^*+b}{2}\right)h^*$. He then multiplied the area of the segment by 3 and added the total area of the segment to the area of the triangle to find the area of the circle. To prove his work, he found the diameter of the circle, then found the area of the circle using the formula $a = \frac{c}{3} \times \frac{c}{4}$ by finding the circumference 'c'; then he added the area of the four pieces to equal the area of the circle.

The Solution

We will consider four pieces of land: the triangle, and the three segments.

We will get the area of the triangle.

Multiply 12 x 12; the result is 144.

Multiply 6 x 6; the result is 36.

Subtract 36 from 144; the result is 108.

Find the square root of 108; the result is $10\,\frac{1}{3}\,\frac{1}{20}\,\frac{1}{120}$ divine cubits.

This is the middle height of the triangle $= h$.

The base is 12 cubits; half of it is 6 cubits.

For the area of the triangle, multiply 6 by $10\,\frac{1}{3}\,\frac{1}{20}\,\frac{1}{120}$, and the result is

$62\,\frac{1}{3}\,\frac{1}{60}$ square cubits.

We shall take $\frac{1}{3}$ of $10\,\frac{1}{3}\,\frac{1}{20}\,\frac{1}{120}$; the result is $3\,\frac{1}{3}\,\frac{1}{10}\,\frac{1}{60}\,\frac{1}{120}\,\frac{1}{180} = h*$.

This is the height of the triangle segment.

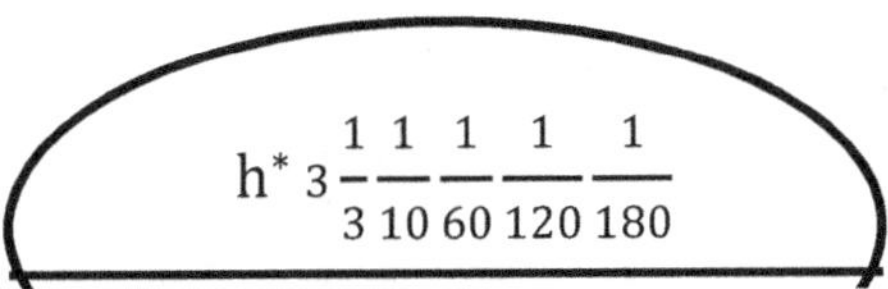

$h* = $ the height of the segment.

We shall add $3\,\frac{1}{3}\,\frac{1}{10}\,\frac{1}{60}\,\frac{1}{120}\,\frac{1}{180}$ to 12;

The result is $15\,\frac{1}{3}\,\frac{1}{10}\,\frac{1}{60}\,\frac{1}{120}\,\frac{1}{180} = \text{h*} + \text{b}$.

Then we take half of it; the result is $7\,\frac{2}{3}\,\frac{1}{20}\,\frac{1}{120}\,\frac{1}{240}\,\frac{1}{360} = \left(\frac{\text{h*}+\text{b}}{2}\right)$.

Then multiply this by $3\,\frac{1}{3}\,\frac{1}{10}\,\frac{1}{60}\,\frac{1}{120}\,\frac{1}{180}$; the result is $26\,\frac{5}{6}\,\frac{1}{10}$ square cubits.

Then we multiply 3 times (the area of the 3 segments); the result is $80\,\frac{2}{3}\,\frac{1}{10}\,\frac{1}{30}$ square cubits.

Add this to $62\,\frac{1}{3}\,\frac{1}{60}$; the result is $143\,\frac{1}{10}\,\frac{1}{20}$ square cubits.

This is the area of the circular plot.

<u>The Proof</u>

The height of the triangle is $10\,\frac{1}{3}\,\frac{1}{20}\,\frac{1}{120}$ cubits.

We add $\frac{1}{3}$ of the height, $3\,\frac{1}{3}\,\frac{1}{10}\,\frac{1}{60}\,\frac{1}{120}\,\frac{1}{180}$; the result is $13\,\frac{5}{6}\,\frac{1}{45} = $ the diameter.

We multiply $13\,\frac{5}{6}\,\frac{1}{45}$ by 3 to get 'C', the circumference.

The result is $41\,\frac{1}{2}\,\frac{1}{15}$ cubits.

$\frac{1}{3}$ of the circumference is $13\,\frac{5}{6}\,\frac{1}{45}$, and $\frac{1}{4}$ of the circumference is $10\,\frac{1}{3}\,\frac{1}{20}\,\frac{1}{120}$.

Multiply $\frac{c}{3} \times \frac{c}{4} = 13\,\frac{5}{6}\,\frac{1}{45}$ by $10\,\frac{1}{3}\,\frac{1}{20}\,\frac{1}{120}$.

The result is $143\,\frac{5}{6}\,\frac{1}{10}\,\frac{1}{30}$ square cubits.

The difference is $\frac{2}{3}\,\frac{1}{10}\,\frac{1}{20}$ square cubits.

<u>The Correct Solution</u>

Area of the triangle $= \frac{1}{2}\mathrm{bh}$

$\mathrm{b} = 12$

$\mathrm{h} = \sqrt{144 - 36} = \sqrt{108} = 10.39304$

The area of the triangle $= \frac{1}{2}(12)(10.39304) = 62.35824$ square cubits.

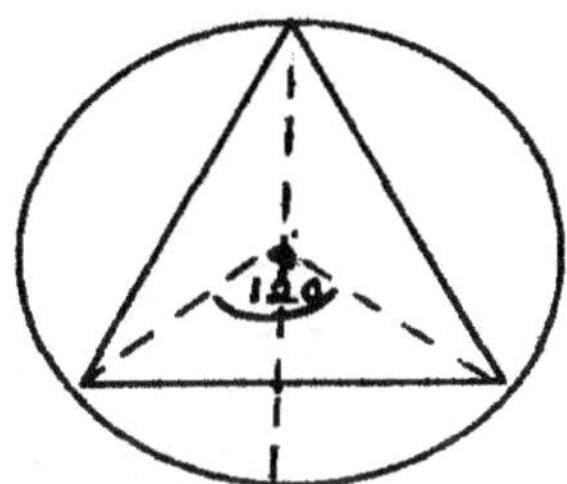

The area of the circular segment $a = \frac{1}{2}\mathrm{r}^2(\theta - \sin\theta)$ where 'r' is the radius

and θ is the angle in radians subtended by the side of the arc of the circle.

$\mathrm{r} = \frac{1}{2}\left(\mathrm{h} + \frac{1}{3}\mathrm{h}\right) = \frac{1}{2}(10.39304) + \frac{1}{3}(10.39304) = 6.928$ cubits.

The area of the segment $\frac{1}{2}\mathrm{r}^2(\theta - \sin\theta)\theta = \frac{2\pi}{3}$ radius

$$\theta = 120 \text{ degrees.}$$

$\frac{1}{2}(6.928)^2\left(\frac{2\pi}{3} - \sin 120\right) = 29.47914$ square cubits.

The area of the segment $= 3 \times 29.4791 = 88.4374$ square cubits.

The total area of the four pieces $= 62.3582 + 88.4374 = 150.7956$ square cubits.

<u>The Proof</u>

The radius of the circle is 6.928

Area of the circle $= \pi r^2 = \pi(6.928)^2$

$$= 150.7876 \text{ square cubits}$$

If we take $\pi = 3$,

then the area $= 3r^2 = 3(6.928)^2 = 143.99$ square cubits.

Problem 37

Given a circular plot of land with an area of 675 square cubits and a diameter of 30 cubits in which is laid out a square, determine the size of the square and the size of the plot.

<u>The Solution</u>

The scribe determined that the diameter of the circle equaled the diagonal of the square which equaled the hypotenuse of the right-angle triangle which is half of the square. He used the Pythagorean Theorem to find the side of the square; then he found the area of the square.

<u>The Proof</u>

To prove that, he got the area of the four circular segments,

$a = \left(\dfrac{h^*+b}{2}\,h^*\right)$, and added it to the area of the square to prove that the total is the area of the circle.

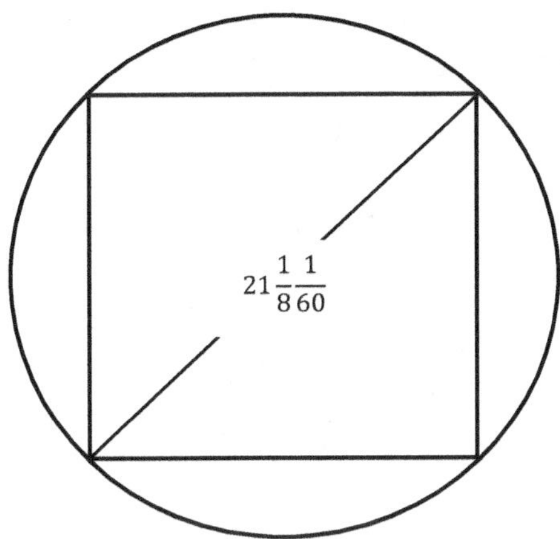

We shall multiply 30 x 30; the result is 900.

Take half of that; the result is 450.

We get the square root*; the result is $21\frac{1}{5}\frac{1}{60}$ divine cubits.

We multiply $21\frac{1}{5}\frac{1}{60}$ by $21\frac{1}{5}\frac{1}{60}$; the result is 450 square cubits.

Note: The square root is rounded off. The correct value should be $21\frac{1}{5}\frac{1}{70}$.

We get the height of the segment.

Take $21\frac{1}{5}\frac{1}{60}$ from 30; the result is $8\frac{2}{3}\frac{1}{10}$.

Take half of that; the result is $4\frac{1}{3}\frac{1}{20}\frac{1}{120}$ h*.

Take the area of the segment,

$$h^* \;=\; 4\frac{1}{3}\frac{1}{20}\frac{1}{120}$$

$$\frac{1}{2}\left(4\frac{1}{3}\frac{1}{20}\frac{1}{120} + 21\frac{1}{5}\frac{1}{60}\right)4\frac{1}{3}\frac{1}{20}\frac{1}{120} \approx 56\frac{1}{4}.$$

Multiply the result by $4 = 225$.

Add $450 + 225 = 675$ square cubits, the area of the plot.

The solution we use now, given the diagonal of the square = 30 with a side of 'x', then:

$$x^2 + x^2 = (30)^2 \quad 2x^2 = 900 \quad a^2 = 450$$

The size of the side $= \sqrt{450} = 21.2132$.

The area of the square $= 21.232 \times 21.2132 \approx 450$ square cubits

$$\frac{1}{2}r^2(\theta - \sin\theta)$$

$$r = \frac{30}{s} = 15 \quad \theta = \frac{\pi}{2}$$

The area of the segment:

$$\frac{1}{2}15^2\left(\frac{\pi}{2} - \sin\frac{\pi}{2}\right) = 64.21 \text{ square cubits.}$$

The square and four segments $= 450 + 4 \times 64.21 = 706.84$ square cubits.

The area of the plot $\pi r^2 = \pi(15)^2 = 706.85$ square cubits.

The difference is using π instead of 3.

Problem 38

Given a circular plot of land in which is laid out an equilateral triangle with sides of 10 cubits each, determine the various areas. Use the area of the segment $a = \left(\frac{h^*+b}{2}\right) h^*$.

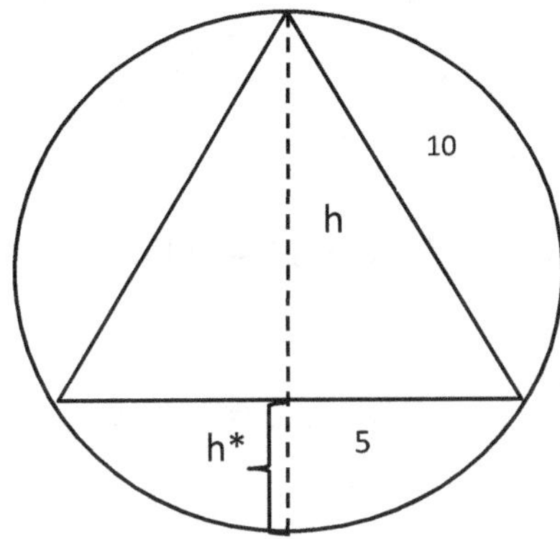

It is similar to Problem 32.

We shall take half of the side, the result is 5.

We multiply $10 \times 10 = 100$ and $5 \times 5 = 25$.

Take 25 from $100 = 75$.

The square root of 75 is $8\frac{2}{3}$.

The area of the triangle $\frac{1}{2} \times 10 \times 8\frac{2}{3} = 43\frac{1}{3}$ square cubits.

Take $\frac{1}{3}$ of $8\frac{2}{3} = 2\frac{5}{6}\frac{1}{30}\frac{1}{45}$.

We shall add $8\frac{2}{3}$ to $2\frac{5}{6}\frac{1}{30}\frac{1}{45}$; the result is $11\frac{1}{2}\frac{1}{30}\frac{1}{45}$.

Multiply 3 times $11\frac{1}{2}\frac{1}{30}\frac{1}{45} = 34\frac{2}{3}$, the circumference of the circle.

$\frac{1}{3}$ of $34\frac{2}{3} = 11\frac{1}{2}\frac{1}{30}\frac{1}{45}$

$\frac{1}{4}$ of $34\frac{2}{3} = 8\frac{2}{3}$

The area of the circle plot is $8\frac{2}{3}$ by $11\frac{1}{2}\frac{1}{30}\frac{1}{45} \approx 100$ square cubits.

We shall get $\frac{1}{2}(h^* + b) = \frac{1}{2}\left(2\frac{5}{6}\frac{1}{30}\frac{1}{45} + 10\right) = \frac{1}{2}\left(12\frac{5}{6}\frac{1}{30}\frac{1}{45}\right) = 6\frac{1}{3}\frac{1}{9}$.

We shall multiply $6\frac{1}{3}\frac{1}{9}$ by $2\frac{5}{6}\frac{1}{30}\frac{1}{45} = h^*$.

The result is $18\frac{1}{2}\frac{1}{10}$; multiply this by 3 which $= 55\frac{5}{6}\frac{1}{60}$ which is the area of the three segments in square cubits.

We shall add this to the area of the triangle $43\frac{1}{3}$; the result is $99\frac{1}{6}\frac{1}{60}$.

The remainder is $\frac{5}{6}\left(\frac{1}{18}\frac{1}{180}\right)$.

Use the Pythagorean Theorem to get the height of the triangle.

$h^2 + 25 = 100$

$h = \sqrt{75} = 8.6605$

The area of the triangle $= \frac{1}{2}(10 \times 8.6605) = 43.3$ square cubits.

The area of the segment is $a = \frac{1}{2}r^2(\theta - \sin\theta)$.

$r = \frac{1}{2}\left(h + \frac{1}{3}h\right) = \frac{1}{2}\left[8.6605 + \frac{1}{3}(8.6605)\right] = 5.7737$ cubits.

$a = \frac{1}{2}(5.7737)^2\left[\frac{1}{3}\pi - \sin\frac{2\pi}{3}\right] = 20.4692$.

The 3 segments $= 3 \times (20.4692) = 61.4077$ square cubits.

Add $61.4077 + 43.3 = 104.7077$ square cubits, the area of the plot.

The difference is the value of π.

Area of the circle plot $\pi r^2 = \pi(5.773)^2 = 104.70$ square cubits.

If we use $\pi = 3$, the answer is 99.98 square cubits.

VOLUME OF A PYRAMID

(Relationship between the Volume, the Height, and the Side of the Base of the Pyramid)

Problem 39 and 40 concern the volume of a square pyramid with $V = \frac{1}{3}hb$, where h = height of the pyramid and b = area of the base which is $b = a^2$, such that 'a' is the side of the base.

Problem 39

Given a pyramid 300 cubits high with a square base 500 cubits to a side, determine the distance from the center of any side to the apex.

The scribe used the Pythagorean Theorem to find the hypotenuse with a side equal to 300 and the other side equal to $\frac{500}{2} = 250$.

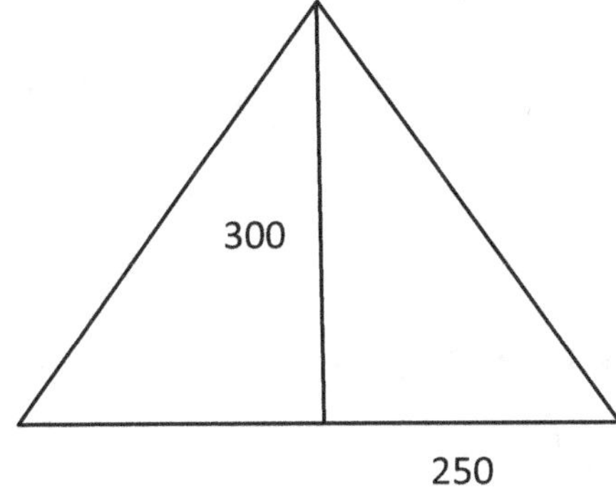

We shall take $\frac{1}{2}$ of the 500; the result is 250 cubits.

We shall multiply 300 by 300; the result is 90,000.

We shall multiply 250 by 250; the result is 62,500.

We shall add 90,000 + 62,500; the result is 152,500.

We will take the square root of $152,500 = 390\frac{1}{2}$ cubits.

The distance from the center to any side of the apex is $390\frac{1}{2}$.

In modern terminology,

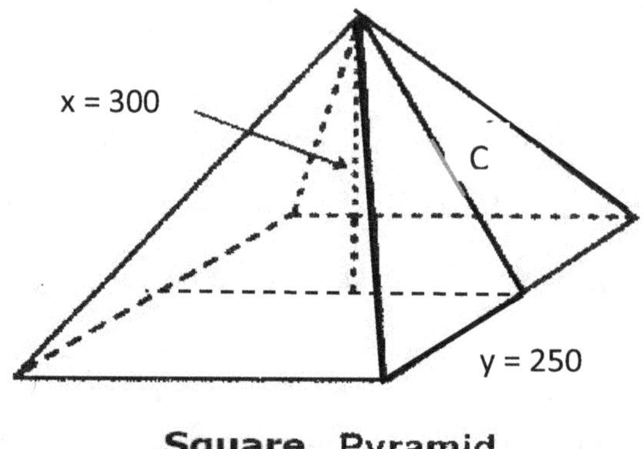

Square Pyramid

Take half of the side 250.

$$x^2 + y^2 = c^2$$

$$(300)^2 + (250)^2 = 152{,}500 = c^2$$

$$c = \sqrt{152{,}500} = 390.5 \text{ cubits.} \quad \text{The accurate value is } 390\frac{20}{39} \text{ cubits.}$$

Problem 40

Given a pyramid with a vertical height of 10 cubits and the square base 10 cubits to a side, determine the volume.

The scribe used the formula that the volume of the pyramid $= \frac{1}{3}$ bh.

'b' is the area of the base. This illustration by the scribe is not correct.* See note.

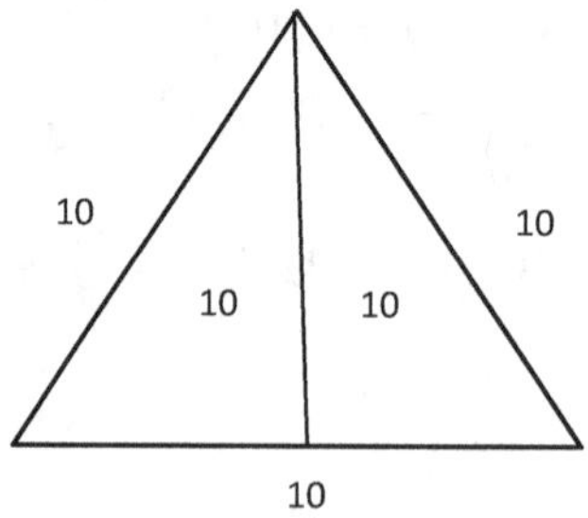

We shall multiply 10 by 10; the result is $100 = b$

We shall multiply 100 x 10; the result is $1{,}000 = bh$

The volume will be $\frac{1}{3}bh$ which is $\frac{1}{3}$ x 1,000 which is $333\frac{1}{3}$ cubic cubits. *

This answer is incorrect.

The corrected illustration:

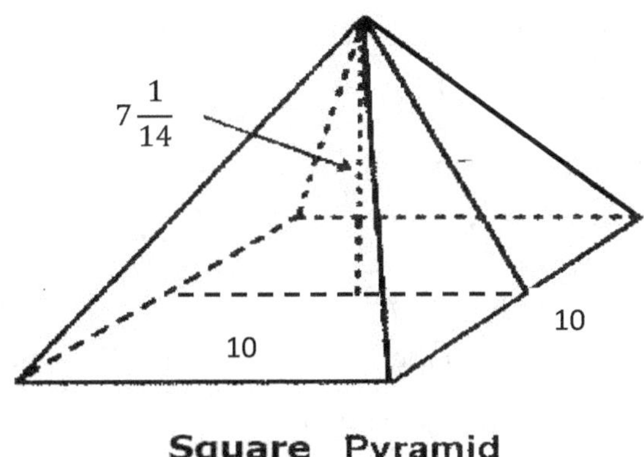

Square Pyramid

Note: If he means the distance of 10 cubits is from the base to the apex along any of the four edges of the pyramid, then the height $= \sqrt{100 - 25} = \sqrt{75} = 7\frac{1}{14}$. The volume would be $\frac{1}{3}\left(10 \text{ x } 10 \text{ x } 7\frac{1}{14}\right) = 235\frac{2}{3}\frac{1}{21}$ cubic cubits, which is not the scribe's answer of $333\frac{1}{3}$ cubic cubits.

Chapter 5

Other Demotic Mathematical Papyri

ROMAN DEMOTIC (LATE PERIOD)

From the beginning of the Roman rule in Egypt, the Demotic was less used in public life. There are, however, a number of literary texts in late Demotic, especially from the first and the second centuries AD. The Demotic Graffite of Isis of Philaer was dated around 452 AD. The scripts in this period resemble the Coptic and the Greek scripts.

Demotic is the second script described on the Rosetta Stone. The Demotic scripts were deciphered before the Hieroglyphic, starting with Silvestre de Sacy. Egyptologists and linguists are known as Demotists. Demotists from the Egyptian Museum believe that some of these papyri were copied from earlier Hieratic or Demotic Papyri.

P. BRITISH MUSEUM 10399 (BMP 10399)

The papyrus was discovered in an unknown place in Egypt and is of the Ptolemaic Period, a later time than the Cairo Papyrus. It is located now in the British Museum as a part of the Robert Hay collection.

Problem 1 has too little preserved to be clear.

The next four problems were identified and all of them are similar.

Problem 2 states that given a mast of a ship of a specific length, with a known diameter at its foot and a known diameter at its top, we are asked to find its volume in hins of water.

The next six problems were that with a given fraction of $\frac{a}{b}$ added to 1, determine what fraction of $1\frac{a}{b}$ must be subtracted from it to give 1 again.

The last problem could not be identified.

P. BRITISH MUSEUM 10520 (BMP 10520)

It is located in the British Museum. 13 mathematic problems were identified in this papyrus as follows:

1. Determine the sum of an arithmetic progression, 1 to 10 and with a constant difference of 1, and its second order. See the series as explained $\sum\limits_{a_1}^{a_n}\sum\limits_{i=1}^{n}i$ on page 180 and 181.

2. Multiplication table of 64 from 1 to 16.

3. Multiply 13 by 17.

4. Resolve $\frac{2}{35}$ to a unit fraction.

5. Multiply $\frac{1}{3}\frac{1}{15}$ by $\frac{2}{3}\frac{1}{21}$.

6. Resolve $\dfrac{\frac{1}{3}\frac{1}{15}}{\frac{2}{3}\frac{1}{21}}$.

7. Multiply $\frac{2}{3}\frac{1}{21}$ by $\frac{1}{3}\frac{1}{15}$.

8. Subtract $\frac{1}{3}\frac{1}{15}$ from $\frac{2}{3}\frac{1}{21}$.

9. Add $\frac{1}{3}\frac{1}{15}$ to $\frac{2}{3}\frac{1}{21}$.

10. Extract the square root of 10.

11. Extract the square root of $\frac{1}{2}$.

12. Given a rectangular piece of land 12 cubits by 10 cubits, determine the area.

13. Given a rectangular field $1\frac{1}{2}$ by $2\frac{1}{4}$ cubits, determine its area.

P. BRITISH MUSEUM 10794 (BMP 10794)

The papyrus was discovered in an unknown place in Egypt and now is located in the British Museum. It contains two multiplication tables; one for $\frac{1}{90}$ from 1 to 10, and the other for $\frac{1}{150}$ from 1 to 10 (Appendix V).

P. CARLSBERG 30

The papyrus is probably from the second century, and it is located in the Carlsberg collection at the University of Copenhagen, originally from Tunis in Fayum. Five problems were identified; one of them may have to do with areas increasing or decreasing in size through a progression in diameters, and the other may be a proportion of gold, copper, and lead; but nothing was preserved to identify the nature of these problems.

AN ANALYSIS OF THE PROBLEMS

P. BRITISH MUSEUM 10399 (BMP 10399)

Problem 2 to Problem 5 in the P. British Museum 10399 have to do with the volume of a ship's mast in hins of water. Since water cannot be poured into a solid mast, the scribe takes the mast as being encased in brass and then being drawn out of the brass case which becomes a receptacle into which the water can be poured. It came out leaving the shape of a truncated cone, or frustum of the cone. In the Mosco Papyrus, 2,000 years before this papyrus, the ancient Egyptian developed the volume for the frustum of the square pyramid with one area having a larger base as 'B,' and the area of the small base as 'A'. The volume of the frustum of the square pyramid

$$= \tfrac{1}{3}h(b^2 + ab + a^2)$$

$$= \tfrac{1}{3}h\left(B + \sqrt{AB} + A\right)$$

such that B, the area of the larger base, $= b^2$,

and A, the area of the smaller base, $= a^2$.

We shall show that the volume of the frustum of the cone will be $\tfrac{1}{3}h\left(B + \sqrt{AB} + A\right)$.

<u>The Proof</u>

Assume that the diameter of the small base is 'a' and of the larger base is 'b,' and the area of the small base is 'A' and the area of the larger base is 'B'.

$$\text{Then } B = \left(b^2 - \tfrac{b^2}{4}\right) = \tfrac{3}{4}b^2, \text{ and } A = \left(a^2 - \tfrac{a^2}{4}\right) = \tfrac{3}{4}a^2 \text{ and } \pi = 3.$$

The volume of the frustum of the cone $= \frac{1}{3}h\left(B + \sqrt{AB} + A\right)$ and 'h' is the height of the frustum.

$h = h_1 + h_2$

The area of the large base is $\frac{3}{4}a^2$ and the area of the small base is $\frac{3}{4}b^2$.

The volume of the frustum is V_T.

Then $V_T = \frac{1}{3}\left(\frac{3}{4}a^2\right)h - \frac{1}{3}\left(\frac{3}{4}b^2\right)h_1 = \frac{1}{4}\left[a^2(h_1 + h_2) - b_1^2 h_1\right]$

$$= \frac{3}{4}\left(\frac{1}{3}a^2 h_1 + \frac{1}{3}a^2 h_2 - \frac{1}{3}b^2 h_1\right)$$

$\frac{h_1}{h} = \frac{b}{a} \rightarrow \frac{h_1}{h_1 + h_2} = \frac{b}{a}; \quad ah_1 = bh_1 + bh_2 \rightarrow ah_1 - bh_1 = bh_2$

$(a - b)h_1 = bh_2 \quad \Rightarrow \quad h_1 = \frac{bh_2}{(a-b)}.$

$$V_T = \frac{1}{3}\left[\left(\frac{3}{4}\right)a^2\left(\frac{bh_2}{(a-b)}\right) + \frac{1}{3}\left(\frac{3}{4}\right)a^2 h_2 - \frac{1}{3}b^2\left(\frac{bh_2}{(a-b)}\right)\right]$$

$$= \left(\frac{1}{3}\right)\left(\frac{3}{4}\right)h_2\left(\frac{a^2 b}{a-b} + a^2 - \frac{b^3}{a-b}\right)$$

$$= \left(\frac{1}{3}\right)\left(\frac{3}{4}\right)h_2\left[\frac{a^2 b + a^2(a-b) - b^3}{a-b}\right]$$

$$= \left(\frac{1}{3}\right)\left(\frac{3}{4}\right)h_2\left(\frac{a^2 b + a^3 - a^2 b - b^3}{a-b}\right)$$

$$= \left(\frac{1}{3}\right)\left(\frac{3}{4}\right)h_2\left(\frac{a^3 - b^3}{a-b}\right) = \left(\frac{1}{3}\right)\left(\frac{3}{4}\right)h_2\left[\frac{(a-b)(a^2 + ab + b^2)}{a-b}\right]$$

$$= \frac{1}{3}h_2\left(\frac{3}{4}a^2 + \frac{3}{4}ab + \frac{3}{4}b^2\right) = \frac{1}{3}h_2\left(A + \sqrt{AB} + B\right)$$

$A = \frac{3}{4}a^2; \quad B = \frac{3}{4}b^2; \quad \sqrt{AB} = \frac{3}{4}ab.$

The scribe used the approximate formula to get the volume of the frustum of the cone. This formula is:

$$V = h_2\left[\frac{3}{4}\left(\frac{a+b}{2}\right)^2\right] \quad \text{for } \pi = 3 \quad a = 2r_1 \quad b = 2r_2$$

$$= \frac{3h_2}{4}\left(\frac{2r_1+2r_2}{2}\right)^2 = \frac{3h_2}{4}(r_1+r_2)^2 = \frac{3}{4}(r_1{}^2+2r_1r_2+r_2{}^2)h_2.$$

To find the error:

$$\frac{h_2}{3}\left(A+\sqrt{AB}+B\right) - \frac{3h_2}{4}(r_1{}^2+2r_1r_2+r_2{}^2) \qquad \pi = 3$$

$$= \frac{h_2}{3}\left(\frac{3}{4}d_1^2+\frac{3}{4}d_1d_2+\frac{3}{4}d^2\right) - \frac{3h_2}{3}(r_1{}^2+2r_1r_2+r_2{}^2)$$

$$= \frac{3h_2}{3}\left(\frac{4r_1{}^2}{4}+\frac{4r_1r_2}{4}+\frac{4r_2{}^2}{4}\right) - \frac{3h_2}{4}(r_1+2r_1r_2+r_2{}^2)$$

$$= h_2(r_1{}^2+r_1r_2+r_2{}^2) - \frac{3h_2}{4}(r_1{}^2+2r_1r_2+r_2{}^2)$$

$$= \frac{h_2}{4}\left(4r_1{}^2+4r_1r_2+4r_2{}^2\right) - \frac{h_2}{4}\left(3r_1{}^2+6r_1r_2+3r_2{}^2\right)$$

$$= \frac{h_2}{4}\left(4r_1{}^2+4r_1r_2+4r_2{}^2-3r_1{}^2-6r_1r_2-3r_2{}^2\right)$$

$$= \frac{h_2}{4}(r_1{}^2-2r_1r_2+r_2{}^2) = \frac{h_2}{4}(r_1-r_2)^2 \qquad \pi = 3$$

for the real value of π.

The error $= \frac{\pi h_2}{12}(r_1-r_2)^2$.

The error becomes smaller when r_2 is closer to r_1.

The error $= 0$ when $r_1 = r_2$.

The error also becomes smaller when h_2 becomes smaller given r_1, r_2 are constant.

Note: $\frac{a+b}{2}$ = average of the 2 diameters.

Problem 1

Too little is preserved in the papyrus so we cannot identify its nature.

Problem 2

Given a mast 100 divine cubits in length, 3 divine cubits in diameter at its foot, and 1 cubit in diameter at its top, determine the volume in hins of water.

The Solution

The scribe used the formula for the area of a circle, $a = \frac{c}{3} \times \frac{c}{4}$.

d is the diameter of the circle, $c = 3d$.

c is the circumference, $a = \frac{3}{4} \times d^2 = d^2 - \frac{d^2}{4}$.

Average of the diameter, $d = \frac{3+1}{2} = 2$ cubits,

$\therefore$ the area of the cross section $= d^2 - \frac{d^2}{4} = (2)^2 - \frac{(2)^2}{4} = 4 - \frac{4}{4} = 3$.

The volume of the mast $=$ the area of the cross section times the length

$$= 3 \times 100 = 300 \text{ cubic cubits.}$$

The hin $= 1$ palm $\times$ 1 palm $\times$ 1 palm $= 1$ cubic palm.

Therefore, the Cubit of Thoth $= 7$ palms.

The number of hins in cubic cubits $= 7 \times 7 \times 7 = 343$ hins of water.

The volume of the mast $= 300 \times 343 = 102,900$ hins of water.

The error $= \frac{h}{4}(r_1 - r_2)^2 = \frac{100}{4}(1.5 - .5)^2 = 26$ $r_1 = 1.5$ $r_2 = .5$

if $\pi = 3$.

The real error $= \frac{\pi h}{12}(r_1 - r_2)^2 = \frac{\pi(100)}{12}(1.5 - .5)^2 = \frac{25\pi}{3} = 8.3\pi$ cubic

cubits.

Problem 3

Given a mast 90 cubits in length, 3 cubits in diameter at its foot, and 1 cubit in diameter at its top, determine its volume in hins of water.

<u>The Solution</u>

The scribe averaged the two diameters d $= \frac{d_1+d_2}{2} = \frac{3+1}{2} = 2$ cubits.

The area of the cross section a $= d^2 - \frac{d^2}{4} = (2)^2 - \frac{(2)^2}{4} = 4\text{-}1 = 3$ sq. cubits.

He found the volume to be 3 x 90 = 270 cubic cubits.

The error $= \frac{\pi h}{12}(r_1 - r_2)^2 = \pi\left(\frac{90}{12}\right)(1.5 - .5)^2 = 7.5\pi$ cubic cubits.

$\qquad r_1 = 1.5 \quad r_2 = .5$

The cubic cubit = 343 hins.

The number of hins for the mast = 270 x 343 = 92,610 hins of water.

Problem 4

Given a mast 80 cubits in length, 3 cubits in diameter at its foot, and 1 cubit in diameter at its top, determine its volume in terms of hins of water.

<u>The Solution</u>

The scribe averaged the two diameters; the result $= \frac{3+1}{2} = 2$ cubits.

Then he found the area of the cross section which is a $= d^2 - \frac{d^2}{4} = 3$ square cubits.

The volume = 3 x 80 = 240 cubic cubits.

The number of hins of the mast = 240 x 343 = 82,320 hins of water.

The error $= \frac{\pi(80)}{12}(1.5 - .5)^2 = 6.7\pi$ cubic cubits.

The error is getting smaller when 'h' is getting smaller.

Problem 5

Given a mast 70 cubits in length, 3 cubits in diameter at its foot, and 1 cubit in diameter at its top, determine its volume in hins of water.

The Solution

The scribe averaged the two diameters, $\frac{3+1}{2} = 2$ cubits. Then he found the area of the cross section; $d^2 - \frac{d^2}{4} = 3$ square cubits. Then he found the volume; $V = 3 \times 70 = 210$ cubic cubits.

The number of hins of the mast $= 210 \times 343 = 72{,}030$ hins of water.

The error $= \frac{\pi h}{12}(r_1 - r_2)^2 = \frac{\pi(70)}{12}(1.5 - .5)^2 = 5.83\pi$ cubic cubits.

This showed that when the height is getting smaller and, given that r_1, r_2 are constant, the error will get smaller.

The papyrus also contains six problems with a general form stated as follows:

We are given the fraction $\frac{a}{b}$ which is added to 1 to determine what fraction of $1\frac{a}{b}$ must be subtracted from it to give 1 again.

The result is $\frac{a}{a+b}$, which the scribe used.

The Proof

$$\left(1 + \frac{a}{b}\right) - p\left(1 + \frac{a}{b}\right) = 1 \qquad \text{'p' is the fraction wanted.}$$

$$\left(1 + \frac{a}{b}\right)(1 - p) = 1$$

$$\left(\frac{b+a}{b}\right)(1 - p) = 1$$

$$1 - p = \frac{b}{a+b}$$

$$p = 1 - \frac{b}{a+b} = \frac{a+b-b}{a+b} = \frac{a}{a+b}.$$

Problem 6, 7, 8, 9, and 10 have a $= 1$ and **Problem 11** has a $= 5$

To prove the result, the scribe used the following:

$$\left(1 - \tfrac{a}{a+b}\right) + \tfrac{a}{b}\left(1 - \tfrac{a}{a+b}\right) = 1.$$

To prove that:

$$\left(1 - \tfrac{a}{a+b}\right) + \tfrac{a}{b}\left(1 - \tfrac{a}{a+b}\right) = \left(1 + \tfrac{a}{b}\right) - \tfrac{a}{a+b}\left(1 + \tfrac{a}{b}\right)$$

$$= \left(1 + \tfrac{a}{b}\right) - \tfrac{a}{a+b}\left(1 + \tfrac{a}{b}\right)$$

$$= 1 + \tfrac{a}{b} - \tfrac{a}{a+b} - \tfrac{a}{b}\left(\tfrac{a}{a+b}\right)$$

$$= \left(1 - \tfrac{a}{a+b}\right) + \tfrac{a}{b} - \tfrac{a}{b}\left(\tfrac{a}{a+b}\right)$$

$$= \left(1 - \tfrac{a}{a+b}\right) + \tfrac{a}{b}\left(1 - \tfrac{a}{a+b}\right)$$

$$= \left(1 - \tfrac{a}{a+b}\right)\left(1 + \tfrac{a}{b}\right)$$

$$= \left(\tfrac{a+b-a}{a+b}\right)\left(\tfrac{b+a}{b}\right) = \left(\tfrac{b}{a+b}\right)\left(\tfrac{a+b}{b}\right) = 1.$$

Problem 6

Given a fraction $\tfrac{1}{5}$ which is added to 1, determine the fraction of $1\tfrac{1}{5}$ that must be subtracted from it to give 1 again.

The Solution

The scribe gives the answer $\tfrac{1}{6} = \tfrac{a}{a+b}$ where a $= 1$ and b $= 5$.

The Proof

$$\left(1 - \tfrac{a}{a+b}\right) + \tfrac{a}{b}\left(1 - \tfrac{a}{a+b}\right)$$

$$\left(1 - \tfrac{1}{6}\right) + \tfrac{1}{5}\left(1 - \tfrac{1}{6}\right) = \tfrac{5}{6} + \tfrac{1}{5}\left(\tfrac{5}{6}\right) = 1.$$

Problem 7

Given the fraction $\frac{1}{6}$ which is added to 1, determine what fraction of $1\frac{1}{6}$ must be subtracted from it to give 1 again.

<u>The Solution</u>

$$a = 1 \quad b = 6$$

The scribe gives the answer $\frac{1}{7}$ or $\frac{a}{a+b}$ $\quad a = 1 \quad b = 6.$

<u>The Proof</u>

He used the identity:

$$\left(1 - \frac{a}{a+b}\right) + \frac{a}{b}\left(1 - \frac{a}{a+b}\right)$$

$$\left(1 - \frac{1}{7}\right) + \frac{1}{6}\left(1 - \frac{1}{7}\right) = \frac{6}{7} + \frac{1}{6}\left(\frac{6}{7}\right) = \frac{6}{7} + \frac{1}{7} = 1.$$

Problem 8

Given the fraction $\frac{1}{7}$ which is added to 1, determine what fraction of $1\frac{1}{7}$ must be subtracted from it to give 1 again.

<u>The Solution</u>

The scribe gives the answer $\frac{1}{8} = \frac{a}{a+b}$ $\quad a = 1 \quad b = 7.$

<u>The Proof</u>

He used the identity:

$$\left(1 - \frac{a}{a+b}\right) + \frac{a}{b}\left(1 - \frac{a}{b}\right); \quad \left(1 - \frac{1}{8}\right) + \frac{1}{7}\left(1 - \frac{1}{8}\right) = \frac{7}{8} + \frac{1}{8} = 1.$$

Problem 9

Given the fraction $\frac{1}{8}$ which is added to 1, determine what fraction of $1\frac{1}{8}$ must be subtracted from it to give 1 again.

The Solution

The scribe gives the answer as $\frac{1}{9} = \frac{a}{a+b}$ $\quad a = 1 \quad b = 8$.

The Proof

The scribe used the identity:

$$\left(1 + \frac{1}{a}\right) - \frac{1}{a+1}\left(1 + \frac{1}{a}\right) = \left(1 - \frac{a}{a+b}\right) + \frac{a}{b}\left(1 - \frac{a}{b}\right)$$

$$\left(1 - \frac{1}{9}\right) + \frac{1}{8}\left(1 - \frac{1}{9}\right) = \frac{8}{9} + \frac{1}{9} = 1.$$

Problem 10

Given the fraction $\frac{1}{9}$ which is added to 1, determine what fraction of $1\frac{1}{9}$ must be subtracted to give 1 again.

The Solution

The scribe gives the answer $\frac{1}{10} = \frac{a}{a+b}$ $\quad a = 1 \quad b = 9$.

The Proof

He used the identity:

$$\left(1 + \frac{1}{a}\right) - \frac{1}{a+1}\left(1 + \frac{1}{a}\right) = \left(1 - \frac{a}{a+b}\right) + \frac{a}{b}\left(1 - \frac{a}{b}\right)$$

$$\left(1 - \frac{1}{10}\right) + \frac{1}{9}\left(1 - \frac{1}{10}\right) = \frac{9}{10} + \frac{1}{10} = 1$$

$$\left(1 + \frac{1}{9}\right) - \frac{1}{10}\left(1 + \frac{1}{9}\right) = 1 - \frac{1}{10} + \frac{1}{10} = 1.$$

Problem 11

Given the fraction $\frac{5}{6}$ which is added to 1, determine what fraction of $1\frac{5}{6}$ must be subtracted to give 1 again.

Note: The scribe wrote $\frac{5}{6}$ as $\frac{5}{6}$, but he could have used it as a unit fraction which is $\frac{1}{3}\frac{1}{11}\frac{1}{33}$. In this problem, there was a new method of representation of the fraction other than as a unit fraction.

The Solution

Using $a = 5$ and $b = 6$, the scribe gives the answer $\frac{a}{a+b} = \frac{5}{11}$.

The Proof

The scribe used the identity:

$$\left(1 - \frac{a}{a+b}\right) + \frac{a}{b}\left(1 - \frac{a}{a+b}\right)$$

$$\left(1 - \frac{5}{11}\right) + \frac{5}{6}\left(1 - \frac{5}{11}\right)$$

$$\frac{6}{11} + \left(\frac{5}{6} \times \frac{6}{11}\right) = \frac{6}{11} + \frac{5}{11} = 1.$$

Problem 12

It is not clear because little remains to determine the details of the problem.

P. BRITISH MUSEUM (10520)

This papyrus contains the next 13 problems.

Problem 1

Determine the sum of the arithmetic progression 1 to 10 and the sum of the same progression 1 to 10 in the second order.

"The first progression is 1, 2, 3, 4, 5, 6, 7, 8, 9, 10".

"The second progression is the first progression of the second order (according to the scribe); 1, 3, 6, 10, 15, 21, 28, 36, 45, 55";

this is the first 10 triangular numbers.

The underlining formula the scribe used for the first progression was:

$$\text{For the first progression } s_n = \sum_{i=1}^{n} i = \frac{n^2+n}{2} \quad \text{and } n = 10$$

The scribe took $n^2 = 10 \times 10 = 100$; then he added $n = 10$ to make 110; he divided by 2 and got 55.

In the second progression, the underlying formula the scribe used was:

$$\text{For the second progression } s_n = \sum_{i=1}^{n} \sum_{i=1}^{n} i = \sum_{i=1}^{n} \frac{(i^2+i)}{2}$$

$$= (\frac{(n+2)}{3}) \frac{(n^2+n)}{2} = \frac{(n+2)}{3} \sum_{i=1}^{n} i$$

*   The Greeks knew the theory of the arithmetic series, but they usually treated it in connection with polygonal numbers. For example, the following are the first five triangular numbers. Each triangular number is $\sum_{1}^{n} n$ (Stewart, 1953, p. 499)

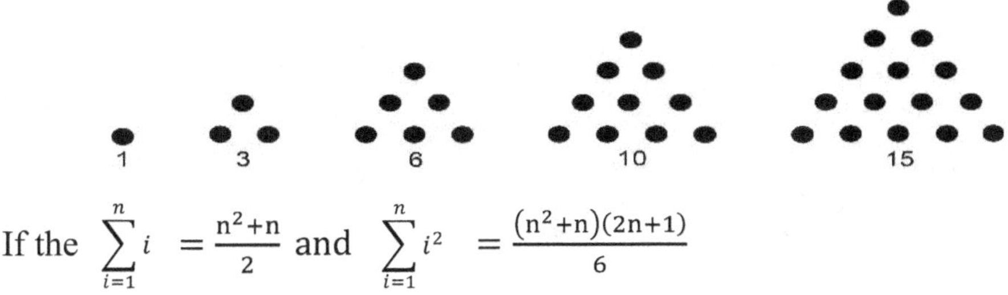

$$\text{If the } \sum_{i=1}^{n} i = \frac{n^2+n}{2} \text{ and } \sum_{i=1}^{n} i^2 = \frac{(n^2+n)(2n+1)}{6}$$

The proof is the following:

$$\sum_{a_1}^{a_n} \sum_{i=1}^{n} i = \left(\frac{n+2}{3}\right) \sum_{i=1}^{n} i$$

$$\sum_{a_1}^{a_n} \left(\frac{i^2+1}{2}\right) = \frac{1}{2}\sum_{i=1}^{n} i^2 + \frac{1}{2}\sum_{n=1}^{n} i = \frac{1}{2}\left[\frac{(n^2+n)(2n+1)}{6} + \frac{(n^2+n)}{2}\right]$$

$$= \frac{(n^2+n)}{2}\left(\frac{2n+1+3}{6}\right) = \frac{(n^2+n)}{2}\frac{(2n+4)}{6} = \frac{(n^2+n)}{2}\frac{(n+2)}{3}$$

$$= \frac{(n+2)}{3}\sum_{i=1}^{n} i$$

The scribe's solution for the second progression used the above formula

with $\sum_{i=1}^{10} i = 55$

and $n + 2 = 12$, then $\frac{1}{3}(n+2) = 4$

then $4 \times 55 = 220 = \frac{1}{3}(n+2)\sum_{i=1}^{10} i$

Problem 2

A multiplication table for 64 from 1 to 16:

1. 64	5. 320	9. 576	13. 832
2. 128	6. 384	10. 640	14. 896
3. 192	7. 448	11. 704	15. 960
4. 256	8. 512	12. 768	16. 1024

Problem 3

Multiply 13 by 17.

The Solution

The scribe used the distributive property of multiplication over addition instead of the usual multiplication methods.

$(a + b)(c + d) = ac + bc + ad + bd.$

$13 \times 17 = (10 + 3)\,10 + (10 + 3)\,7.$

Multiply $10 \times 10 = 100$; then $10 \times 3 = 30$.

Multiply $7 \times 10 = 70$; and $3 \times 7 = 21$.

Then he added $100 + 30 + 70 + 21$ for a result of 221.

Problem 4

Resolve $\dfrac{2}{35}$ to its unit fraction.

The Solution

The scribe used the expression: If $n = pg$ and $p + g$ is divisible by 2,

then $\dfrac{2}{n} = \dfrac{1}{g\left(\frac{p+g}{2}\right)} + \dfrac{1}{p\left(\frac{p+g}{2}\right)}$

The scribe factored $35 = 5 \times 7 = p \times g \qquad p = 5$ and $g = 7$

He added $5 + 7 = 12$; then $\dfrac{5+7}{2} = 6$

$\dfrac{2}{35} = \dfrac{1}{5\times6} + \dfrac{1}{7\times6} = \dfrac{1}{30} + \dfrac{1}{42}.$

Problem 5

Multiply $\frac{1}{3}\frac{1}{15}$ by $\frac{2}{3}\frac{1}{21}$.

The scribe used the expression: If $(1 + g) = np$, and if $(2g + 1) = np$,

then $\frac{1}{p} + \frac{1}{pg} = \frac{n}{g}$.

The Proof

$$\frac{2}{p} + \frac{1}{pg} = \frac{2g+1}{pg} = \frac{np}{pg} = \frac{n}{g}.$$

In general: if $kg + 1 = np$, then $\frac{k}{p} + \frac{1}{pg} = \frac{n}{g}$.

The Proof

$$\frac{k}{p} + \frac{1}{pg} = \frac{(kg+1)}{pg} = \frac{np}{pg} = \frac{n}{g}.$$

The Solution

The scribe's solution stated that we shall multiply 5 x 7; the result is 35. If he applied the rule for the fraction, it will be 5 and the second fraction will be 7.

$\frac{2}{3} + \frac{1}{21}$ to the number 7 will result in 5 or $\frac{5}{7}$.

We shall bring $\frac{1}{3}\frac{1}{15}$ to the number 5. The result is 2 or $\frac{2}{5}$. Either he got this from the table or he was familiar with the previous rule.

$p = 3 \quad g = 5 \quad k = 1 \quad kg + 1 = 6 \quad n = 2$

$$\frac{1}{p} + \frac{1}{pg} = \frac{n}{g}$$

$\frac{1}{3} + \frac{1}{15} = \frac{2}{5}$; he called this two parts of 5.

If p = 3 and g = 7, then 2g + 1 = 15.

$$n = \frac{15}{3} = 5 \quad \text{If } (2g + 1) = np, \text{ then } \frac{2}{p} + \frac{1}{pg} = \frac{(2g+1)}{pg} = \frac{np}{pg} = \frac{n}{g}.$$

He then added $\frac{2}{3} + \frac{1}{21} = \frac{5}{7}$ and called this 5 parts of 7.

He multiplied 2 x 5 = 10 and 5 x 7 = 35.

This is 10 parts of 35 $= \left(\frac{10}{35} = \frac{40}{140} = \frac{35}{140} + \frac{5}{140} = \frac{1}{4}\frac{1}{28} \right)$

The result $\frac{1}{4}\frac{1}{28}$, (which equals $\frac{2}{7}$), probably is from the table.

Problem 6

Resolve $\dfrac{\frac{1}{3}\frac{1}{15}}{\frac{2}{3}\frac{1}{21}}$.

The Solution

The scribe knew $\frac{1}{3}\frac{1}{15} = 2$ parts of 5, and $\frac{2}{3}\frac{1}{21}$ is 5 parts of 7;

then he took 5 x 7 = 35. (He was finding a common denominator.)

$\frac{2}{3}\frac{1}{21} = \frac{5}{7}$ of 35 = 25

$\frac{1}{3}\frac{1}{15} = \frac{2}{5}$ of 35 = 14

$\frac{14}{35} \div \frac{25}{35} = \frac{14}{25}$; then make the case that 14 makes a part of 25.

$\left(\frac{14}{25} \right) = \frac{12\frac{1}{2}}{25} + \frac{1}{25} + \frac{\frac{1}{2}}{25}.$

The result is $\frac{1}{2}\frac{1}{25}\frac{1}{50}.$

The scribe may have used the rule $\frac{a}{n} \div \frac{b}{n} = \frac{a}{b}.$

Problem 7

Similar to Problem 5, except instead of asking to multiply $\frac{2}{3}\frac{1}{21}$ times $\frac{1}{3}\frac{1}{15}$, he asks to take $\frac{1}{3}\frac{1}{15}$ of $\frac{2}{3}\frac{1}{21}$. The scribe used the same solution as in Problem 5.

Problem 8

Subtract $\frac{1}{3}\frac{1}{15}$ from $\frac{2}{3}\frac{1}{21}$.

The Solution

The same methods for subtracting two fractions are as follows:

The scribe used 5 x 7 = 35 as a common denominator.

Find $\frac{2}{3}\frac{1}{21}$ of 35; the result is 25.

Find $\frac{1}{3}\frac{1}{15}$ of 35; the result is 14.

Subtract 14 from 25; the result is 11.

Take 11 parts from 35; the result is $\frac{1}{4}\frac{1}{28}\frac{1}{35}$, which is the remainder.

Problem 9

Add $\frac{1}{3}\frac{1}{15}$ to $\frac{2}{3}\frac{1}{21}$.

The Solution

The scribe used the same method as that for the addition of fractions.

The scribe used 5 x 7 = 35.

Find $\frac{2}{3}\frac{1}{21}$ of 35; the result is 25.

Find $\frac{1}{3}\frac{1}{15}$ of 35; the result is 14.

We shall add 25 + 14; the result is 39.

Then we carry 35 into 39; the result is $\left(1\frac{8}{70}\right) = 1\frac{1}{10}\frac{1}{70}$.

The next two problems are to find the square root.

The formula of approximate square roots was known to the Egyptians in Problem 1 in the Berlin Papyrus (~1325 BC) and the Kahun Papyrus (~1750 BC) – a thousand years before Heron of Alexandria – where it was found in his work in the second half of the first century AD, according to Neugebauer (1957, p. 178).

He used the formula $\sqrt{n} = \sqrt{a^2 \pm b} = a \pm \dfrac{b}{2a}$.

Problem 10

Find the square root of 10.

The Solution

We shall take 3; 3 times 3 is $9 = a^2$

The remainder is 1, then $\sqrt{10} = \sqrt{9+1}$ $\qquad\qquad a = 3 \quad b = 1$

We shall multiply $\dfrac{1}{2}$ by $\dfrac{1}{3}$ $= \dfrac{1}{6} = \dfrac{b}{2a}$

We shall add $\dfrac{1}{6}$ to 3; the result is $3\dfrac{1}{6} = a + \dfrac{b}{2a}$

$3\dfrac{1}{6}$ is the square root.

The Proof

Multiply $3\dfrac{1}{6} \times 3\dfrac{1}{6}$, the result is $10\dfrac{1}{36}$.

The difference in the square root is $\dfrac{1}{36}$.

Problem 11

Extract the square root of $\frac{1}{2}$.

The Solution

The scribe chose $\frac{1}{2}$ to be $\frac{18}{36}$.

Then he used the formula $\sqrt{a^2 + b} = a + \frac{b}{2a}$

$$\sqrt{18} = \sqrt{16 + 2} = 4 + \frac{2}{2(4)} = 4\frac{1}{4}.$$

He knew that 6 x 6 = 36; then $\sqrt{\frac{18}{36}} = \frac{4\frac{1}{4}}{6}$.

We shall bring $4\frac{1}{4}$ to the number $\frac{1}{4}$;

the result is $17 \left(4\frac{1}{4} = \frac{17}{4} \right)$.

We shall bring 6 to $\frac{1}{4}$;

the result is $24 \left(6 = \frac{24}{4} \right)$.

We know 17 parts of 24 is $\left(\frac{17}{24} \right) = \left(\frac{16}{24} + \frac{1}{24} \right)$;

the result is $\frac{2}{3}\frac{1}{24}$.

The Proof

Reckon $\left(\frac{2}{3}\frac{1}{24} \right) \times \left(\frac{2}{3}\frac{1}{24} \right)$.

The result is $\frac{1}{2}\frac{1}{576}$.

The difference of the squared root is $\frac{1}{576}$.

The next two problems are to determine the area of rectangular land.

Problem 12

Given a piece of land 12 cubits by 10 cubits, determine the area.

<u>The Solution</u>

We shall add south and north.

The result is 20; its half is 10.

We shall add east and west.

The result is 24; its half is 12.

We shall reckon 10 × 12.

The result = 120 square cubits.

We shall carry 100 into 120.

The result is $1\frac{1}{5}$ land cubits (a land cubit = 100 square cubits).

***Note*:**

The scribe knew the area of the rectangle, and he could then multiply 10 x 12 to get 120. He often used the following formula incorrectly when dealing with a four-sided area of *unequal* lengths. It would only have been correctly used for an area of a rectangle.

$$Area = \left(\frac{south + north}{2}\right)\left(\frac{east + west}{2}\right)$$

Problem 13

Given a field $1\frac{1}{2}$ by $2\frac{1}{4}$, determine the area. The scribe gives the dimensions in arouras. He means the dimensions to be khats which is 100 cubits. Then the area will be in arouras. The aroura $= 100 \times 100 = 10,000$ square cubits.

The Solution

We shall add the east and west; the result is $4\frac{1}{2}$; its half is $2\frac{1}{4}$.

We shall add north and south; the result is 3; its half is $1\frac{1}{2}$.

We shall reckon $2\frac{1}{4} \times 1\frac{1}{2}$; the result is $3\frac{1}{4}\frac{1}{8}$ auroras.

P BRITISH MUSEUM 10794

Problem 1

A multiplication table for $\dfrac{1}{90}$ from 1 to 10.

*The final four fractions are preserved; the rest is a possible restoration according to the trend:

1. $\dfrac{1}{90} = \dfrac{1}{90}$ (1)

2. $\dfrac{2}{90} = \dfrac{1}{45}$ (2)

3. $\dfrac{3}{90} = \dfrac{1}{30}$ (3)

4. $\dfrac{4}{90} = \left(\dfrac{1}{30}\dfrac{1}{90}\right)$ (3+1)

5. $\dfrac{5}{90} = \left(\dfrac{1}{30}\dfrac{1}{45}\right)$ (3+2)

6. $\dfrac{6}{90} = \left(\dfrac{1}{15}\right)$ (6)

7. $*\dfrac{7}{90} = \left(\dfrac{1}{15}\dfrac{1}{90}\right)$ (6+1)

8. $*\dfrac{8}{90} = \left(\dfrac{1}{15}\dfrac{1}{45}\right)$ (6+2)

9. $*\dfrac{9}{90} = \dfrac{1}{10}$ (9)

10. $*\dfrac{10}{90} = \left(\dfrac{1}{10}\dfrac{1}{90}\right)$ (9+1)

The scribe can generate numbers 4, 5, 7, 8, and 10 from numbers 1, 2, 3, 6, and 9.

Problem 2

A multiplication table of $\frac{1}{150}$ from 1 to 10.

1. $\frac{1}{150} = \frac{1}{150}$ (1)

2. $\frac{2}{150} = \frac{1}{90}\frac{1}{450}$ (2)

3. $\frac{3}{150} = \frac{1}{60}\frac{1}{300}$ (3)

4. $\frac{4}{150} = \frac{1}{45}\frac{1}{225}$ (4)

5. $\frac{5}{150} = \frac{1}{30}$ (5)

6. $\frac{6}{150} = \frac{1}{30}\frac{1}{150}$ (5,1)

7. $\frac{7}{150} = \frac{1}{30}\frac{1}{90}\frac{1}{450}$ (5,2)

8. $\frac{8}{150} = \frac{1}{20}\frac{1}{300} = \left(\frac{1}{30}\frac{1}{60}\frac{1}{300}\right)$ (5,3)

9. $\frac{9}{150} = \frac{1}{30}\frac{1}{45}\frac{1}{225}$ (5,4)

10. $\frac{10}{150} = \frac{1}{15}$ (10)

The scribe can generate 6, 7, 8, and 9, from 1, 2, 3, 4, 5, and 10.

Note: $\frac{8}{150} = \frac{1}{30}\frac{1}{60}\frac{1}{300}$; the scribe used $\frac{1}{30} + \frac{1}{60} = \frac{3}{60} = \frac{1}{20}$ and wrote $\frac{8}{150} = \frac{1}{20}\frac{1}{300}$.

P. CARLSBERG 30

Problem 1

Too little is preserved to identify its nature.

Problem 2

These three figures and two lines are preserved.

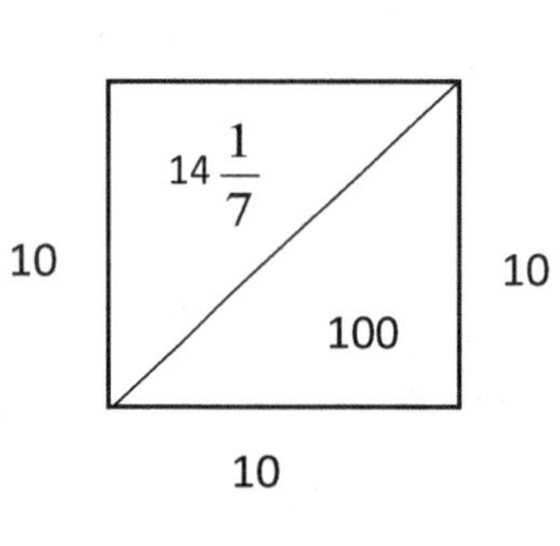

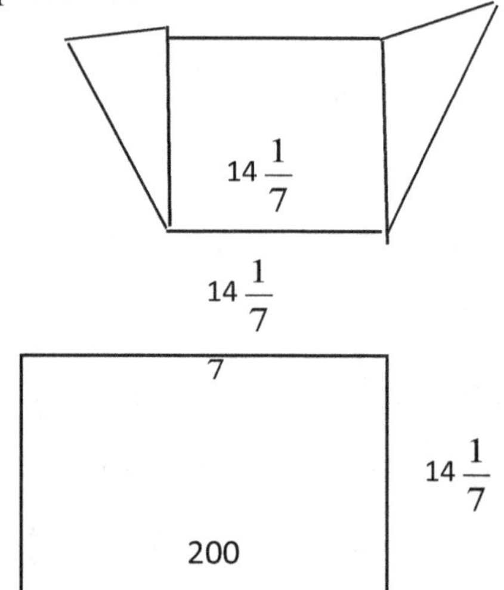

We may speculate that the scribe tried to show how to construct a square with an area twice of a given square using the diagonal of the given square as a side of the new square so the new square will be double the area of the first square; or the area increasing or decreasing in size through a progression in the dimensions; or another method of doubling the square by adding the two triangles with a base equal to $\frac{1}{2}$ the given base of the square and the height is the same height of the squares.

Problem 3

Too little was preserved to identify the nature of the problem.

Problem 4

Only the end of the fifth line is preserved, and we cannot identify the nature of the problem.

Problem 5

It has the proportions of gold, silver, copper, and lead; but there is nothing preserved to identify the nature of the problem.

Note: **Problem 62** in the Rhind Mathematical Papyrus states that a bag containing equal weights of gold, silver, and lead has been bought for 84 shaty. What is the amount in the bag of each precious metal if given a deben of gold being 12 shaty, for a deben of silver is 6 shaty, and for a deben of lead is 3 shaty.

<u>The Solution</u> (of the scribe)

Add that which is given for a deben of each precious metal.

$$12 + 6 + 3 = 21$$

Multiply 21 to get 84 shaty, for which this bag was bought.

The result is 4 which is the number of deben in each precious metal.

Do the following:

12 x 4 = 48 shaty for the gold in the bag.

6 x 4 = 24 shaty for the silver in the bag.

3 x 4 = 12 shaty for the lead in the bag.

21 x 4 = 84 shaty for the total cost of the bag.

Note: The shaty was a seal or coin and here represents a unit of value instead of weight. The papyrus does not say that the bag contains equal weights of gold, silver and lead, but in the solution the scribe proceeds as if this condition was understood.

Chapter 6

A Comparison between Hieratic & Demotic Papyri

In the Hieratic Papyri, we can distinguish two types of texts: problem (or procedural) texts and table texts. Egyptian mathematical problem texts begin by stating a mathematical problem and then providing step-by-step instructions for solving the problem. We can differentiate several sections of a mathematical problem text.

1. The title is written and the method of calculating a problem is often highlighted by the use of red ink.
2. The given data in the problem is often followed by a question that asks what shall be determined. The data are concrete values.
3. The procedure to solve the problem is spelled out in step-by-step instructions using the arithmetic concrete data. Each instruction usually is comprised of one arithmetic operation, such as addition, subtraction, multiplication, division, doubling, etc. After each step, we find an intermediate result provided for a technical analysis of the procedure found in the hieratic text.
4. The indication of the solution usually equals the result of the last step of instruction.
5. Verification of the solution.
6. The problems teach basic mathematical techniques, such as the arithmetic operation, and $\frac{2}{3}$ of any Egyptian fraction.

7. Table texts comprise mathematical information in table format. The tables include those for fractions, most notably the $\frac{2}{n}$ table preserved at the beginning of the Rhind Papyrus. See other tables in Appendix V.

8. The multiplication in hieratic text was often expressed in writing, which rendered multiplication tables unnecessary.

The Demotic Mathematical Papyri show some similarities with the Hieratic. There are problems or procedures, texts, table texts, and mathematical techniques that are expressed rhetorically using specific numerical examples. On the other hand, there are some differences that show improvement on their mathematical knowledge in the area of arithmetic which we will show part of in the comparison between Hieratic Papyri and Demotic Papyri.

PROBLEM TEXTS

Multiplication by whole numbers was confined to 2 or multiples of 2 and, in some cases, it was multiplication by 10 and multiples of 10. Then they added these products from multiplication that will make up the given multiplier. Also in the fractions, they used $\frac{1}{2}$ and $\frac{2}{3}$ and $\frac{1}{10}$, and they doubled $\frac{1}{10}$ to get $\frac{1}{5}$ and halved $\frac{2}{3}$ to get $\frac{1}{3}$. The reciprocals of other numbers were sometimes used as multipliers. They often used the reciprocal of a number, multiplied by the number itself, to give the result of 1. Also, they used the rule that if $ab = c$, then $\left(\frac{1}{c}\right) b = \frac{1}{a}$; in this case, any fraction can be an intermediate multiplier.

MULTIPLICATION

To multiply 13 x 17 in the Hieratic Papyrus, the following methods can be used:

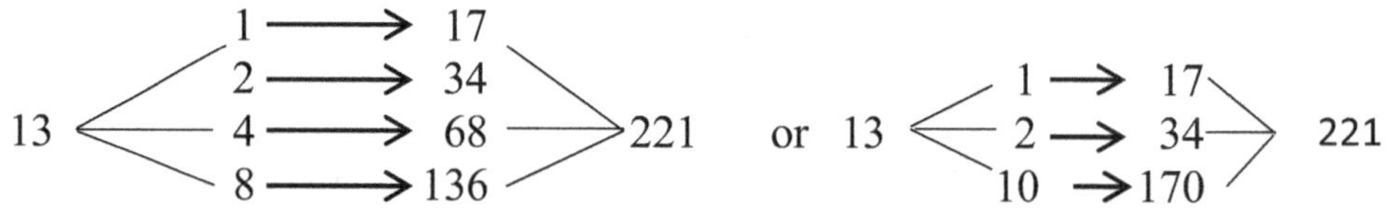

$$13 \begin{cases} 1 \longrightarrow 17 \\ 2 \longrightarrow 34 \\ 4 \longrightarrow 68 \\ 8 \longrightarrow 136 \end{cases} 221 \qquad \text{or} \qquad 13 \begin{cases} 1 \longrightarrow 17 \\ 2 \longrightarrow 34 \\ 10 \longrightarrow 170 \end{cases} 221$$

The result is 221.

In **Problem 3 (PBM 10520)** *(Demotic Papyrus)*, the scribe used another alternate method of multiplying 13 x 17 by using
$13 = 10 + 3$ and $17 = 10 + 7$.
$13 \times 17 = (10 + 3)\,10 + (10 + 3)\,7 = 100 + 30 + 70 + 21 = 221$.

Thus, these multiplication cases are solved by using the **Distributive Property of Multiplication over Addition**.

In the Demotic Papyri, we find that the operation of multiplication is seldom presented in its additive steps, except when fractions are involved. For example, in **Problem 5** of the Cairo Papyrus, the scribe multiplied 17 by 30, and the result was 510, without details.

In **Problem 3** of the Cairo Papyrus, the scribe multiplied $6\frac{18}{47}$ by $15\frac{2}{3}$.

<u>The Solution</u>

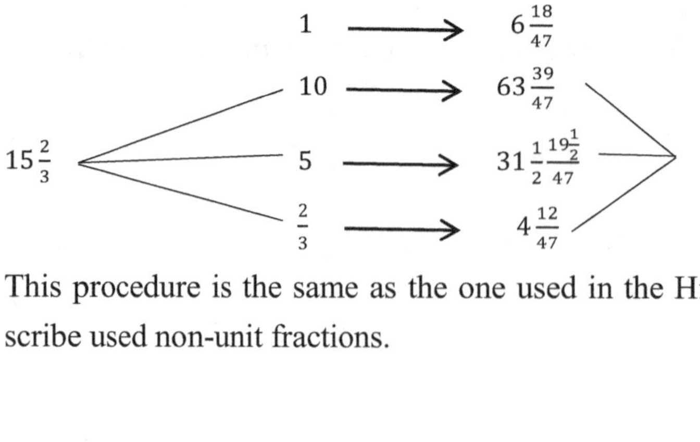

$$15\frac{2}{3} \begin{cases} 1 \longrightarrow 6\frac{18}{47} \\ 10 \longrightarrow 63\frac{39}{47} \\ 5 \longrightarrow 31\frac{1}{2}\frac{19\frac{1}{2}}{47} \\ \frac{2}{3} \longrightarrow 4\frac{12}{47} \end{cases}$$

$$63\frac{39}{47} + 31\frac{1}{2}\frac{19\frac{1}{2}}{47} + 4\frac{12}{47}$$
$$= 99\frac{1}{2}\frac{23\frac{1}{2}}{47} = 100$$

This procedure is the same as the one used in the Hieratic Papyri, except the scribe used non-unit fractions.

DIVISION

The process of division is very close to the process of multiplication. If we are to divide 'a' by 'b', then we try to find 'c' such that b x c = a.

Problem 66 (RMP) *(Hieratic Papyrus)*

If 3200 ro of fat is given out for a year, what is the amount given out in a day? (Divide 3200 by 365). The scribe tells us to do this for problems of this kind.

The Solution

The scribe tried to find the number 'c' which, if multiplied by 365, gives the answer of 3200 as follows:

$$1 \longrightarrow 365$$
$$2 \longrightarrow 730$$
$$4 \longrightarrow 1460$$
$$8 \longrightarrow 2920 \qquad 16(365) = 2(2920) > 3200$$

We stop doubling

$$8\frac{2}{3}\frac{1}{10}\frac{1}{2190}$$

$$\frac{2}{3} \longrightarrow 243\frac{1}{3}$$
$$\frac{1}{10} \longrightarrow 36\frac{1}{2}$$
$$\frac{1}{2190} \longrightarrow \frac{1}{6} \qquad \text{(since 6 x 365 = 2190)}$$

The answer is $8\frac{2}{3}\frac{1}{10}\frac{1}{2190}$ ro of fat is given out per day.

The Proof

$$2920 + 243\frac{1}{3} + 36\frac{1}{2} + \frac{1}{6} = 3200 \text{ ro of fat}$$

In the solution of **Problem 2** in the Cairo Papyrus, it states that $100 \div 53 = 1\frac{47}{53}$ without any detail.

Problem 2 (Cairo Papyrus) *(Demotic Papyrus)*

Divide 100 by $17\frac{2}{3}$.

The Solution

Multiply $17\frac{2}{3}$ by 3.   The scribe chose $6 \times 17\frac{2}{3} = 106$.   He estimated the answer at less than 6 (similar to long division).

Multiply $17\frac{2}{3}$ by $3 = 53$

$100 \div 53 = 1 + \frac{47}{53}$, using a fraction with a numerator greater than 1.

Then multiply the result by $3\left(1 + \frac{47}{53}\right) = 5\frac{35}{53}$, with which the scribe is familiar (a numerator greater than 1).

FROM THE $\frac{2}{n}$ TABLE (RMP) *(HIERATIC PAPYRUS)*

Divide 2 by 35.

The Solution

$$\frac{1}{30}\,\frac{1}{42} \ \Bigg< \ \begin{matrix} \frac{1}{30} \longrightarrow 1\frac{1}{6} \\[2mm] \frac{1}{42} \longrightarrow 2\frac{1}{3}\frac{1}{6} \end{matrix} \ \Bigg> \ 2$$

The answer is $\frac{1}{30}\,\frac{1}{42}$.

Problem 4 (PBM 10520) (*Demotic Papyrus*)

Divide 2 by 35.

The Solution

Multiply 5 by 7; the result $= 35$ where $p = 5$ and $g = 7$

Add 5 and 7; the result $= 12$ where this is $p + g$

Divide 12 by 2; the result $= 6$ where this is $\dfrac{p+g}{2}$

Take 5×6; the result $= 30$ where this is $p\left(\dfrac{p+g}{2}\right)$

Take 7×6; the result $= 42$ where this is $g\left(\dfrac{p+g}{2}\right)$

The answer is $\dfrac{1}{30}\dfrac{1}{42}$ where this is $\dfrac{1}{p\left(\frac{p+g}{2}\right)}\dfrac{1}{g\left(\frac{p+g}{2}\right)}$

$\therefore$ this shows that the scribe applied a general rule to solve the problem, which can be solved by normal division.

In the Demotic Papyrus the scribe, in his solution, applied the general rule that $\dfrac{2}{pg} = \dfrac{1}{p\left(\frac{p+g}{2}\right)} + \dfrac{1}{g\left(\frac{p+g}{2}\right)}$ if $p + g$ is divisable by 2.

FROM THE $\frac{2}{n}$ TABLE (RMP) *(HIERATIC PAPYRUS)*

Divide 2 by 83.

The Solution

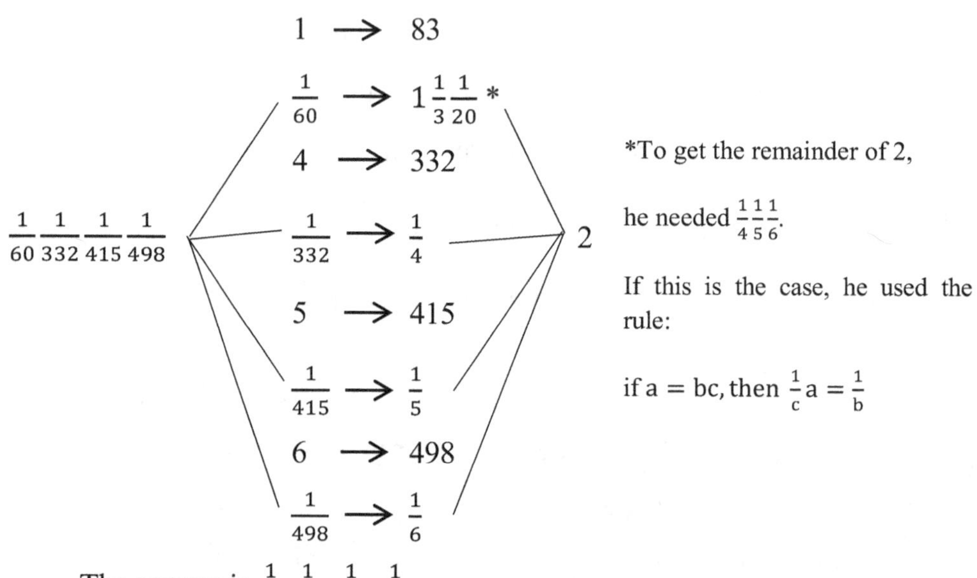

The answer is $\frac{1}{60}\ \frac{1}{332}\ \frac{1}{415}\ \frac{1}{498}$.

By using this method, Ahmes developed the $\frac{2}{n}$ Table that contains a division of 2 by fifty odd numbers from 3 to 101.

THE EGYPTIAN FRACTIONS

The Egyptian fraction was believed to have a numerator equal to 1. In the Cairo Papyrus, in Problems 2, 3, 10, and 13, and in Problem 11 of the P. British Museum 10399, there are fractions with numerators greater than 1. In writing these fractions, the numerator was written first and underlined, such as in Problem 3, and the denominator followed the same line. In a later papyrus, the denominator was underlined, such as in Problem 5 in P. Carlsberg 30. In both cases, the numerator was first. In Problem 2, the answer was $5\frac{35}{53}$, and 35 was underlined.

Once they chose the denominator, it remained the same throughout the solution of the problem, even though this may result in mixed numerators. In Problem 3 in the Cairo Papyrus, for example, the scribe used $6\frac{18}{47}, \frac{19\frac{1}{2}}{47},$ and $\frac{11\frac{1}{2}}{47}$. In Problem 13, even though the problem is not clear, we observe $8\frac{1}{3}\frac{1}{3}\frac{1}{15}\frac{929\frac{1}{210}}{131}$, an odd mixture of proper and improper fractions, with 131 as a denominator occurring in this problem.

ADDITION OF FRACTIONS

An example is to add the fractions $\frac{1}{3}\frac{1}{4}$ and $\frac{1}{6}\frac{1}{12}$.

To add the two groups, the scribe used the highest denominator as a common denominator, but not necessarily the <u>least</u> common denominator. For example, he will choose 12, then use the value of each fraction as part of 12. Then he adds all the values of the parts and the result will be all the parts divided by 12:

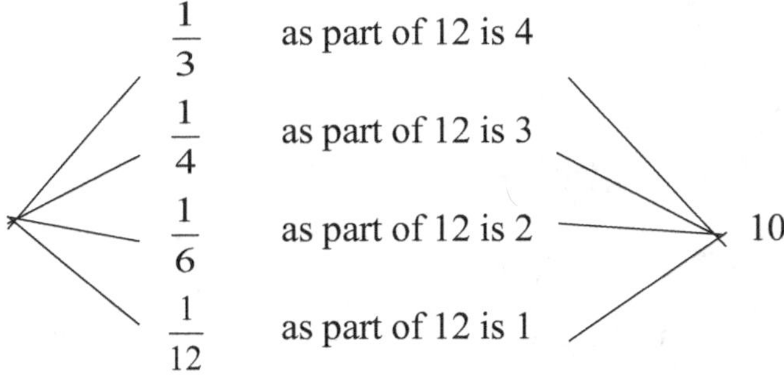

To complete the problem he will try to find the number if he multiplies by 12 for a result of 10.

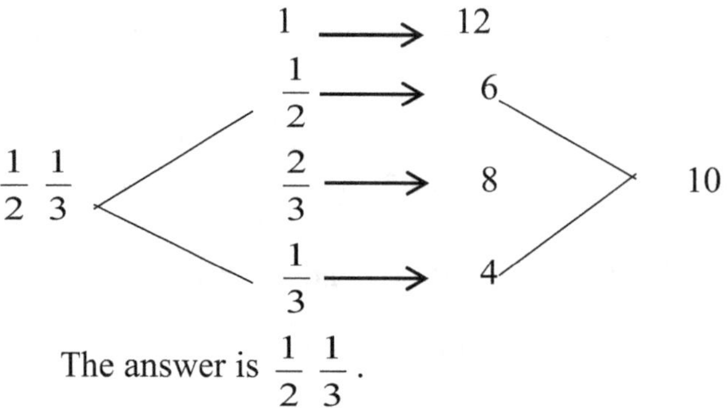

The answer is $\frac{1}{2}$ $\frac{1}{3}$.

Problem 9 (PBM 10520) *(Demotic Papyrus)*

This problem asks that we add $\frac{1}{3}\frac{1}{15}$ to $\frac{2}{3}\frac{1}{21}$.

<u>The Solution</u>

$15 = 3 \times 5$ and $21 = 3 \times 7$

$5 \times 7 = 35$

$\frac{2}{3}\frac{1}{21} = \frac{5}{7}$ as a part of 35 is 25

$\frac{1}{3}\frac{1}{15} = \frac{2}{5}$ as part of 35 is 14

We shall add $25 + 14$; the result is 39

We carry 35 into 39; the result is $1\frac{1}{10}\frac{1}{70}$.

The scribe added every two fractions to a single fraction; then he took $5 \times 7 = 35$ to be the least common denominator.

Subtraction of Fractions

Problem 21 (RMP) *(Hieratic Papyrus)*

The problem states that we complete $\frac{2}{3}\frac{1}{15}$ to 1. (Example of the three steps of the **Process of Completion.**)

$$1 - \frac{2}{3}\frac{1}{15}$$

1. Multiply 15 by $\frac{2}{3}\frac{1}{15}$

2. The result is 10 and $1 = 10 + 1 = 11$

3. The remainder is 4 to get 4 out of 15

$$1 \longrightarrow 15$$
$$10 \longrightarrow 1\frac{1}{2}$$

$$\frac{1}{5}\frac{1}{15} \Big\langle \begin{array}{l} \frac{1}{5} \longrightarrow 3 \\[4pt] \frac{1}{15} \longrightarrow 1 \end{array} \Big\rangle \; 4$$

$\frac{1}{5}\frac{1}{15}$ is what needs to be added to $\frac{2}{3}\frac{1}{15}$ to get 1.

Problem 4 (CMP) *(Demotic Papyrus)*

Given the series $\frac{2}{3}, \frac{1}{6}, \frac{1}{60}, \frac{1}{120}$, and $\frac{1}{210}$, determine the remainder.

The scribe subtracted $\frac{1}{210} - \frac{1}{120}$.

He factored $210 = 7 \times 30$ ab

and $120 = 4 \times 30$ cb

Subtract 4 from $7 = 3$ c – a

Divide 4 by 3

$$1\frac{1}{3} \Big\langle \begin{array}{l} 1 \longrightarrow 3 \\[4pt] \frac{1}{3} \longrightarrow 1 \end{array} \Big\rangle \; 4 \qquad\qquad \frac{c-a}{a}$$

Multiply $210 \times 1\frac{1}{3} = 280$.

The remainder is $\frac{1}{280}$.

The scribe may have applied the formula:

$$x = ab \quad \text{and } y = cb; \quad \text{then } \frac{1}{x} - \frac{1}{y} = \left(\frac{c-a}{a}\right)\frac{1}{y}.$$

MULTIPLICATION OF FRACTIONS

Problem 10 (RMP) *(Hieratic Papyrus)*

Multiply $\frac{1}{4}\frac{1}{28}$ by $1\frac{1}{2}\frac{1}{4}$.

<u>The Solution</u>

$$1 \rightarrow \frac{1}{4}\frac{1}{28} \qquad \frac{1}{4} \text{ of 28 is 7 and } \frac{1}{28} \text{ of 28 is 1}$$

$$\frac{1}{2} \rightarrow \frac{1}{8}\frac{1}{56} \qquad \frac{1}{8} \text{ of 28 is } 3\frac{1}{2} \text{ and } \frac{1}{56} \text{ of 28 is } \frac{1}{2}$$

$$\frac{1}{4} \rightarrow \frac{1}{16}\frac{1}{112} \qquad \frac{1}{16} \text{ of 28 is } 1\frac{3}{4} \text{ and } \frac{1}{112} \text{ of 28 is } \frac{1}{4}$$

$$\text{Total} = 7 + 1 + 3\frac{1}{2} + \frac{1}{2} + 1\frac{3}{4} + \frac{1}{4} = 14$$

$$14 \text{ out of } 28 = \frac{1}{2}$$

The answer is $\frac{1}{2}$

Problem 5 (BMP 10520) (*Demotic Papyrus*)

Multiply $\frac{1}{3}\frac{1}{15}$ by $\frac{2}{3}\frac{1}{21}$.

<u>The Solution</u>

The scribe's solution stated that we shall multiply 5 x 7; the result is 35. If he applied the rule for the fraction, it will be 5 and the second fraction will be 7.

We shall bring $\frac{2}{3}+\frac{1}{21}$ to the number 7 resulting in 5 or $\left(\frac{5}{7}\right)$.

We shall bring $\frac{1}{3}\frac{1}{15}$ to the number 5 resulting in 2 or $\left(\frac{2}{7}\right)$. Either he got this from the table, or he was familiar with the rule $\frac{1}{p}+\frac{1}{pg}=\frac{n}{g}$.

$p = 3 \quad y = 5 \quad k = 1 \quad kg + 1 = 6 \quad n = 2$

$$\frac{1}{p}+\frac{1}{pg}=\frac{n}{g}$$

$\frac{1}{3}+\frac{1}{15}=\frac{2}{5}$; he called this two parts of 5.

If $p = 3$ and $g = 7$, then $2g + 1 = 15$.

$n = \frac{15}{3} = 5$ if $(2g+1) = np$, then $\frac{2}{p}+\frac{1}{pg}=\frac{(2g+1)}{pg}=\frac{np}{pg}=\frac{n}{g}$.

He then added $\frac{2}{3}+\frac{1}{21}=\frac{5}{7}$ and called this 5 parts of 7.

He multiplied 2 x 5 = 10 and 5 x 7 = 35.

This is 10 parts of 35 $= \left(\frac{10}{35}=\frac{40}{140}=\frac{35}{140}+\frac{5}{140}=\frac{1}{4}+\frac{1}{28}\right)$ using scaling.

The result $\frac{1}{4}\frac{1}{28}$, (which equals $\frac{2}{7}$), probably is from the table.

DIVISION OF FRACTIONS

Problem 7 (RMP) *(Hieratic Papyrus)*

Divide $\frac{1}{2}$ by $\frac{1}{4}\frac{1}{28}$.

The Solution

The scribe will find a number which, if he multiplied by $\frac{1}{4}\frac{1}{28}$, would result in $\frac{1}{2}$ as follows:

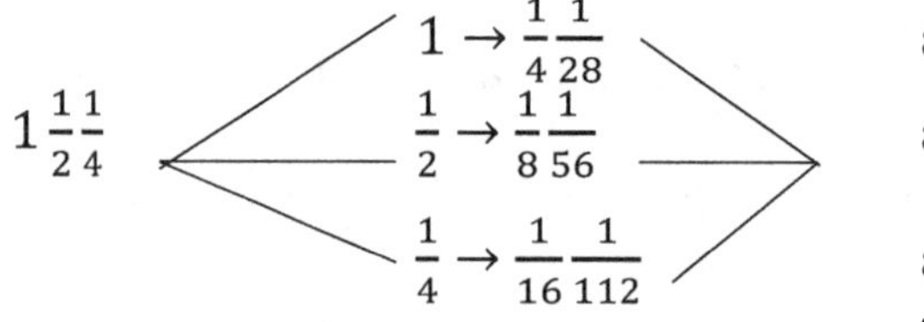

as a part of 112 is 32

as a part of 112 is 16

as a part of 112 is 8

Total as part of 112 is 56.

The answer is $1\frac{1}{2}\frac{1}{4}$.

Problem 6 (BMP 10520) *(Demotic Papyrus)*

Divide $\frac{1}{3}\frac{1}{15}$ by $\frac{2}{3}\frac{1}{21}$.

The Solution

The scribe knew $\frac{1}{3}\frac{1}{15} = 2$ parts of 5, and $\frac{2}{3}\frac{1}{21}$ is 5 parts of 7.

He took $5 \times 7 = 35$. (He was finding a common denominator.)

$$\frac{2}{3}\frac{1}{21} = \frac{5}{7} \text{ of } 35 = 25$$

$$\frac{1}{3}\frac{1}{15} = \frac{2}{5} \text{ of } 35 = 14$$

$$\frac{14}{35} \div \frac{25}{35} = \frac{14}{25}\,;\text{ then make the case that 14 makes a part of 25.}$$

$$\frac{\frac{14}{35}}{\frac{25}{35}} = \frac{14}{25}$$

$$\frac{14}{25} = \frac{12\frac{1}{2}}{25} + \frac{1}{25} + \frac{\frac{1}{2}}{25}.$$

The result is $\frac{1}{2}\frac{1}{25}\frac{1}{50}$.

The scribe may have used the rule if $\dfrac{a}{n} \div \dfrac{b}{n} = \dfrac{a}{b}$.

SQUARE ROOT

In the Hieratic Papyri, the square root of a number is always stated, but not calculated. In the Mosco Papyrus, Struve's translation of Problem 6 is that the square root of 16 is stated as 4, and in the Berlin Papyrus, the square root of $1\frac{1}{2}\frac{1}{4}$ is stated as $1\frac{1}{4}$. These could have been read from a table of the squares of integers. Similar tables for fractions could be constructed.

In the Demotic Papyrus, Problem 10 (PM 10520), the scribe found the square root by applying a formula $\sqrt{n} = \sqrt{a^2 \pm b} = a \pm \frac{b}{2a}$ to find that

$$\sqrt{10} = \sqrt{9+1} = 3 + \frac{1}{6} = 3\frac{1}{6}.$$

Problem 11 (BMP 10520) *(Demotic Papyrus)*

Find $\sqrt{\frac{1}{2}}$.

The scribe converted $\frac{1}{2}$ to an equal fraction with a denominator having a rational square root. In this case, he chose the denominator to be 36. He converted $\frac{1}{2}$ to $\frac{18}{36}$; then he used the rule $\sqrt{\frac{a}{b}} = \frac{\sqrt{a}}{\sqrt{b}} = \frac{\sqrt{ab}}{b}$. Scale $\frac{1}{2}$ to be $\frac{18}{36}$

In this case, $\sqrt{\frac{18}{36}} = \frac{\sqrt{18}}{\sqrt{36}} = \frac{\sqrt{18}}{6} = \frac{\sqrt{16+2}}{6} = \frac{4 + \frac{2}{2\times4}}{6}$.

Using the formula $\sqrt{a^2 + b} = a + \frac{b}{2a}$; then $\frac{4\frac{1}{4}}{6} = \frac{4}{6}\frac{1}{24} = \frac{2}{3}\frac{1}{24}$.

The operations of addition and subtraction are similar. Multiplication and division are built on the theorem that **any positive integer can be expressed by the sum of some terms of the sequence of 2^n for n = 0, 1, 2, ..., n** (doubling and adding).

The calculation in the operation indicated in the procedure was written in the text in detail in the Hieratic mathematics. In Demotic mathematics, the calculation of the operation is seldom written in the text.

The fractions in Hieratic mathematics are expressed as unit fractions or by the Horus Eye. In Demotic mathematics, **the fraction was a unit fraction and proper fraction which are an odd mixture of proper and improper fractions.** The numerator was written first and underlined, and the denominator followed the same line. In later papyri, the denominator was underlined in both cases with the numerator first. Also, any fraction can be represented as an Egyptian fraction.

In addition and subtraction, the Hieratic and Demotic are similar as they used common denominators, but not necessarily the least common denominator. Usually, the highest denominator of the fractions was used and would be added. Once the denominator was chosen, it remained throughout the solution of the problem, even though it may result in mixed numerators. In the solution of the addition of fractions, we find the scribes applied some mathematical rules for addition of the fraction, which was explained and proved in the previous chapter.

The operations of multiplication and division of the fractions in the procedure were written step-by-step in the text format (Hieratic Papyri). In the Demotic Papyri, the details of the operation indicated in the procedure seldom were written in the text.

In the Hieratic Papyri, the square of numbers, both integral and fractional, are stated and calculated, but the square roots are stated and not calculated. This could be read from the Table of Square Integers.

In the Demotic Papyri, the square root was approximated by the rule of $\sqrt{a^2 + b} = a + \frac{b}{2a}$ within the written text. This suggested the scribes were familiar with the rules of multiplication and division (Appendix V).

TABLE TEXTS

The EMLR Table is good evidence for the existence of tables for the addition and subtraction of fractions. In Appendix IV is the complete table with the modern mathematical analysis.

In the Rhind Mathematical Papyrus, there is the table of division of 2 by odd numbers from 3 to 101 after the scribe performed all the division in detail. There is also a table of division of the numbers 1-9 by 10. The scribe performs the division of 1, 2, 3, 4, 5, 6, 7, 8, and 9 by 10. The table was also written in the Reisner Mathematical Papyrus which allowed the builders to construct the different chambers of the temple. This table can be used to take $\frac{1}{10}$ of any number which is not a multiple of 10.

Problem 61 (RMP) *(Hieratic Papyrus)* was a table of multiplication of the fractions (see Appendix V).

Problem 81 (RMP) *(Hieratic Papyrus)* is a table to express the fraction of a hekat in terms of hinu.

The following are examples of how the scribe constructed the $\frac{2}{n}$ table where n = 3 to 101 and the $\frac{n}{10}$ table where n = 1 to 9.

<u>Example 1</u> (RMP) *(Hieratic Papyrus)*

Divide $\frac{2}{5}$.

The Solution

The scribe used the division procedures as follows:

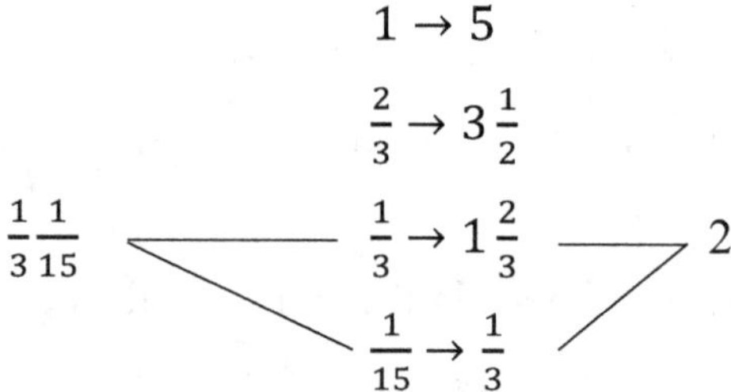

The answer is $\frac{1}{3}\frac{1}{15}$.

Example 2 (RMP) *(Hieratic Papyrus)*

Divide 8 by 10.

The Solution

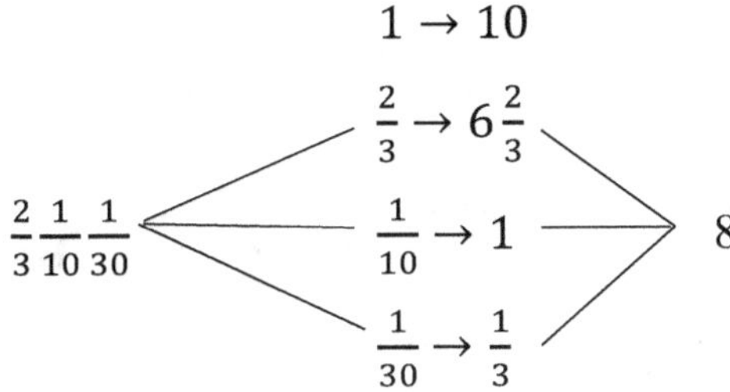

The answer by the scribe is $\frac{2}{3}\,\frac{1}{10}\,\frac{1}{30}$.

The Proof

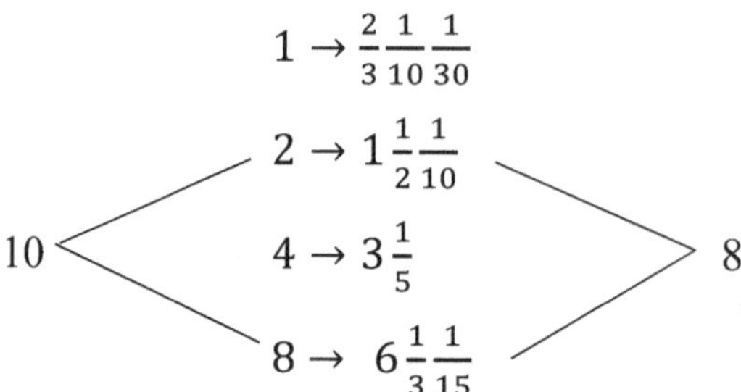

In the Demotic Papyri are two tables of multiplication. **Problem 1** in the **BMP 10794** is a multiplication table of $\frac{1}{90}$ from 1 to 10.

(1) $\quad \frac{1}{90} = \frac{1}{90}$ $\qquad\qquad\qquad\qquad\qquad\qquad \frac{1}{90}$ (1)

(2) $\quad \frac{2}{90} = \frac{1}{45}$ $\qquad$ The scribe divided $\frac{90}{2} = 45$ and $\frac{2}{90} = \frac{1}{45}$ (2)

(3) $\quad \frac{3}{90} = \frac{1}{30}$ $\qquad$ The scribe divided $\frac{90}{3} = 30$ and $\frac{3}{90} = \frac{1}{30}$ (3)

(4) $\quad \frac{4}{90} = \frac{1}{30}\frac{1}{90}$ $\qquad = \frac{3}{90} + \frac{1}{90} = \frac{1}{30} + \frac{1}{90}$ (3+1)

(5) $\quad \frac{5}{90} = \frac{1}{30}\frac{1}{45}$ $\qquad = \frac{3}{90} + \frac{2}{90} = \frac{1}{30} + \frac{1}{45}$ (3+2)

(6) $\quad \frac{6}{90} = \frac{1}{15}$ $\qquad$ The scribe divided $\frac{90}{6} = 15$ and $\frac{6}{90} = \frac{1}{15}$ (6)

(7) $\quad \frac{7}{90} = \frac{1}{15}\frac{1}{90}$ $\qquad = \frac{6}{90} + \frac{1}{90} = \frac{1}{15}\frac{1}{90}$ (6+1)

(8) $\quad \frac{8}{90} = \frac{1}{15}\frac{1}{45}$ $\qquad = \frac{6}{90} + \frac{2}{90} = \frac{1}{15}\frac{1}{45}$ (6+2)

(9) $\quad \frac{9}{90} = \frac{1}{10}$ $\qquad = \frac{6}{90} + \frac{3}{90} = \frac{1}{10}$ (9)

(10) $\quad \frac{10}{90} = \frac{1}{10}\frac{1}{90}$ $\qquad = \frac{1}{10} + \frac{1}{90}$ (9+1)

Number (4) is adding (3) + (1), and number (5) is adding (3) + (2). Number (7) is adding (6) + (1) and number (8) is adding (6) + (2). Lastly, number (10) is adding (9) + (1).

Problem 2 (BMP 10794) (*Demotic Papyrus*)

Multiplication Table for $\frac{1}{150}$ from $1 - 10$.

(1) $\quad \dfrac{1}{150} = \dfrac{1}{150}$ $\hspace{4cm}$ (1)

(2) $\quad \dfrac{2}{150} = \left(\dfrac{6}{450} = \dfrac{5}{450} + \dfrac{1}{450}\right) = \dfrac{1}{90}\dfrac{1}{450}$ $\hspace{1cm}$ (2)

(3) $\quad \dfrac{3}{150} = \left(\dfrac{6}{300} = \dfrac{5}{300} + \dfrac{1}{300}\right) = \dfrac{1}{60}\dfrac{1}{300}$ $\hspace{1cm}$ (3)

(4) $\quad \dfrac{4}{150} = \dfrac{1}{45}\dfrac{1}{225}$ $\hspace{3cm}$ (4) Multiply (2) by 2

(5) $\quad \dfrac{5}{150} = \dfrac{1}{30}$ $\hspace{4.5cm}$ (5)

(6) $\quad \dfrac{6}{150} = \dfrac{1}{30}\dfrac{1}{150}$ $\hspace{3.5cm}$ (5 + 1)

(7) $\quad \dfrac{7}{150} = \dfrac{1}{30}\dfrac{1}{90}\dfrac{1}{450}$ $\hspace{2.8cm}$ (5 + 2)

(8) $\quad \dfrac{8}{150} = \left(\dfrac{16}{300} = \dfrac{15}{300} + \dfrac{1}{300}\right) = \dfrac{1}{20}\dfrac{1}{300}*$ $\hspace{0.5cm}$ (5 + 3)

(9) $\quad \dfrac{9}{150} = \dfrac{1}{30}\dfrac{1}{45}\dfrac{1}{225}$ $\hspace{2.8cm}$ (5 + 4)

(10) $\dfrac{10}{150} = \dfrac{1}{15}$ $\hspace{4cm}$ (10)

$* \dfrac{1}{20} = \dfrac{1}{30}\dfrac{1}{60}$

The preceding tables of multiplication and division are different from the Hieratic Papyri tables. In the Hieratic Papyri, items in the table were found by the standard multiplication and addition procedures; sometimes there was a proof of the results.

In the Demotic Papyri, items in the table were developed by using some of the previous items to find the next items utilizing addition operations.

Chapter 7

Commentary on the Egyptian Arithmetic

Elementary arithmetic was based on three kinds of numbers: whole numbers, fractions, and rational and irrational numbers. There were six operations: four of these were addition, subtraction, multiplication, and division. The fifth operation comprised multiplying a number by itself given a number of times, and the sixth operation was the process in reverse – that of finding a number multiplied by itself given a number of times equal to some given number.

The ancient Egyptian numerals were written in either of three systems of writing: hieroglyphic, hieratic, and demotic. The hieroglyphic was used mainly on monuments of stone, wood, or metal. A more cursive writing script was developed when other media was used, particularly papyrus. It resulted from a more rapid use of reed pens and represented a more rounded form. This form was hieratic and later evolved into a more abbreviated style known as demotic.

Numeration was in decimals like our own system today, but the repetitive principle of hieroglyphic numeration has been replaced by the introduction of ciphers, or special signs, to represent the digits and multiples of the power of 10. Six was not written as a six stroke but, in hieratic, as

$\mathcal{Z}$, as shown on page 9. **"The principle of cipherization, introduced by the Egyptians 4,000 years ago and used by Ahmes in the Rhind Mathematical Papyrus (RMP), represented an important contribution to numeration, and it is one of the factors that makes our system in use today the effective instrument that it is"** (Boyer & Merzbach, 1989, p. 15).

In Egyptian mathematics, the old belief was that a unit fraction had a numerator of 1, and these numbers were distinguished from the integers by having a dot in hieratic or the sign, $\bigcirc$, in hieroglyphic. However, in the demotic texts, as in problems 2, 3, 10, and 13, numbers were mixed in with unit fractions or others with numerators greater than one.

The numerator is written first, and the denominator follows on the same line. In problem 2, 3, 10, and 13 in the Cairo Papyrus, the numerator is underlined. In the later problems, the denominator was underlined. In any one problem the denominator, once chosen, stays the same throughout, even though it may result in a mixed number as a numerator. In problem 3 of the Cairo Papyrus, we see that along with $\frac{6}{47}$, we have $\frac{18}{47}$, $\frac{19\frac{1}{2}}{47}$, and $\frac{11\frac{1}{2}}{47}$. In problem 13 of the Cairo Papyrus, there is a mixture of numbers and improper fractions. From this, the Egyptian later wrote in the Demotic Papyri the unit fraction, the proper fraction, and the improper fraction.

Special symbols were used for $\frac{1}{2}$ and $\frac{2}{3}$, and parts of the Horus Eye were used to represent the fractions $\frac{1}{2}$, $\frac{1}{4}$, $\frac{1}{8}$, $\frac{1}{16}$, $\frac{1}{32}$, and $\frac{1}{64}$. The Egyptians used the supposition that **every positive rational fraction can be represented by the sum of some of the Egyptian fractions.**

The Egyptians developed tables for unit fractions. In the Ahmes Papyrus, there are tables for the expansion of unit fractions. In the EMLR, a table which contains 24 Egyptian fractions, they convert to not-so-elegant Egyptian fractions.

From these tables, we discover some observations which make the expansion elegant.

1. The smaller of the terms of expansion were preferred and, in the EMLR tables, were never more than four terms.
2. The unit fractions are always set down in descending order of the values of their fractions, and never is the same fraction used twice.
3. The smallest and the first denominator is the main consideration, but the scribe accepts larger first denominators if it will greatly reduce the last number.
4. Even numbers are preferred to odd numbers. Although the odd numbers might be larger and the number of terms might be increased, the scribe sought the simpler values available in order to make them more convenient for calculation.

Many attempts were made to discover the methods the Egyptians used to calculate the table. Most of these methods were described as algebraic identities. However, they did not match any single identity which suggested that they knew several methods to represent the rational numbers used in modern times. Some of these methods were suggested and explained in Chapter 2.

The Egyptians used irrational numbers in solving some of the problems, but they used a series of rational numbers to approximate the irrational number. They used the rational numbers as an approximation, rather than use distinct groups of numbers.

ADDITION AND SUBTRACTION

The arithmetic was essentially additive. To add to a number, collect all similar symbols and replace every ten of them by the next higher order of symbols. In one example from the Ahmes Papyrus, addition and subtraction signs were represented by using figures resembling legs of a person advancing for addition ⋀ and departing for subtraction ⋀. Subtraction was a reversal of the process of addition. (See page 11.) If necessary, the Egyptian replaced the higher order by ten of one order below. Most of the time, in the problems which needed addition or subtraction, the scribes gave the answer without showing how the operation was performed.

The Egyptian system was based on symbols, and any number could be written as a collection of these symbols. No difficulties arose from having a zero or place holder in this system. Adding or cutting symbols made the addition and subtraction in the system quick and easy.

MULTIPLICATION

The foundation for the multiplication procedure was built on the rule that **any integer can uniquely be expressed as the sum of selected terms of the series 2^n, where n = 0, 1, 2, 4, 8, ...** . The rule can be proved by inductive reasoning. The confidence with which the Egyptians approached all forms of multiplication suggested that they were aware of this general rule.

When the ancient Egyptian needed to multiply two numbers, he would multiply one of the numbers by repeatedly using the process of doubling (multiplying by 2), adding up all of the selected intermediate multipliers until they added up to the original multipliers. This essentially was a form of binary arithmetic. This process of multiplication requires prior knowledge of only addition and multiples of 2 (doubling). There was a table for 2^n.

This method was used with some modification by the Greeks up until the Middle Ages in Europe. Also, it is popular now in the rural communities of Egypt, the Near East, Ethiopia, and also Russia. With this method, they avoided the need to learn multiplication tables. Also, as a short cut for larger numbers, the multiplicand could be multiplied by 10, 100, etc.

Some mathematicians describe these methods as awkward because it is different from the way we do it now where we multiply by unit figures of our multiplier and then by ten figures, and so on, and then add the result. The ancient method is much closer to the way the binary computer works for multiplication and division.

When the Egyptians multiplied two numbers, they were familiar with the commutative property of addition and multiplication within the number system. They also used the distributive property in multiplication over addition, such as $13 \times 17 = (10 + 3)10 + (10 + 3)7$

$$= 100 + 30 + 70 + 21 = 221.$$

DIVISION

The process of division was close to the process of multiplication. If the scribe wanted to divide 'X' by 'Y', then he wanted to find the number which is multiplied by 'Y' to find 'X'. He would start with 'Y'. How many times should he add 'X' to itself to get 'Y'? Successive multiplication of the divisor was required until the dividend was obtained.

ADDITION OF FRACTIONS

The scribe used the tables, such as the EMLR, $\frac{2}{n}$, and $\frac{n}{10}$ tables, to add some specific fractions. We also find some rules which were used to add specific fractions.

In general, if the scribe wanted to add two groups of fractions, he used methods to find a common denominator, but not necessarily the least common denominator (LCD). In some problems the scribe used the LCD, but scribes might often use a denominator that was smaller or larger than the LCD. If we used these methods, then some of the numerators of the fractions would not be integers after changing the denominator to the common denominator selected for this case.

Since the Egyptian fractions are a unit fraction and $\frac{2}{3}$, the scribe used the common denominators to be suitable to the fractions he added, and if the result had fractions, they would be Egyptian fractions.

Problem 19 (RMP)

Add $\frac{1}{12} + \frac{1}{18} + \frac{1}{36}$.

The scribe chose 18 as a common denominator, which is less than the LCD of 36. He would solve the problem as follows:

$\frac{1}{12}$ is $1\frac{1}{2}$ parts of 18

$\frac{1}{18}$ is 1 part of 18

$\frac{1}{36}$ is $\frac{1}{2}$ parts of 18

Total: 3 is $\frac{1}{6}$ part of 18   (the result is $\frac{1}{6}$).

In our solution, we would use the LCD of 36.

$$\frac{1}{12} + \frac{1}{18} + \frac{1}{36} = \frac{3}{36} + \frac{2}{36} + \frac{1}{36} = \frac{6}{36} = \frac{1}{6}$$

Problem 8 (RMP)

Add $\frac{1}{4} + \frac{1}{6} + \frac{1}{12}$.

The scribe chose a common denominator of 18, which is greater than the LCD of 12.

$\frac{1}{4}$ is $4\frac{1}{2}$ parts of 18

$\frac{1}{6}$ is 3 parts of 18

$\frac{1}{12}$ is $1\frac{1}{2}$ parts of 18

Total: 9 is $\frac{1}{2}$ part of 18 (the result is $\frac{1}{2}$).

In our solution, we would use the LCD of 12.

$$\frac{1}{4} + \frac{1}{6} + \frac{1}{12} = \frac{3}{12} + \frac{2}{12} + \frac{1}{12} = \frac{6}{12} = \frac{1}{2}.$$

It should be noted that the numerators of the fractions to be added are unit fractions, and if the result has a fraction, it also is a unit fraction.

Some of the answers of the problems for addition of fractions are written without details, which suggests that they found them from tables or that they used algebraic identities equivalent to the following identities:

1) $\frac{1}{b} + \frac{1}{b(ab-1)} = \frac{a}{ab-1}$

2) $\frac{1}{a} + \frac{1}{ka} = \frac{(k+1)}{ka}$

3) If 'p' is small, odd, and prime, then: $\frac{2}{p+1} + \frac{2}{p(p+1)} = \frac{2}{p}$

4) $\frac{1}{a} + \frac{(2a-p)}{ap} = \frac{2}{p}$ where 'a' is a practical number $\frac{p}{2} < a < p$

5) $\frac{1}{aq} + \frac{1}{apq} = \frac{2}{pq}$ if $a = \frac{p+1}{2}$

6) If $g + 1 = np$, then $\frac{1}{p} + \frac{1}{pg} = \frac{n}{g}$

7) If $2g + 1 = np$, then $\frac{2}{p} + \frac{1}{pg} = \frac{n}{g}$

8) If $kg + 1 = np$, then $\frac{k}{p} + \frac{1}{pg} = \frac{n}{g}$

SUBTRACTION OF FRACTIONS

If the scribes had addition tables, then they could use the tables for $\frac{1}{a} + \frac{1}{b} = \frac{1}{c}$

to get $\frac{1}{c} - \frac{1}{a} = \frac{1}{b}$ and $\frac{1}{c} - \frac{1}{b} = \frac{1}{a}$. Also, they could use the other addition

identities above and use the same procedure. Other identities were used such

as:

$$\text{If } a = kb \text{ and } b = (k-1)m, \text{ then } \frac{1}{b} - \frac{1}{a} = \frac{1}{km}.$$

In general, the scribe used the common denominator to subtract the fractions.

MULTIPLICATION OF FRACTIONS

The ancient Egyptian used tables for multiplication, such as the table in

Problem 61 (RMP). From the tables and the solution of the problems in the

papyri, they applied the following rules:

1) $\frac{1}{a} \times \frac{1}{b} = \frac{1}{ab}$.

2) $\frac{2}{3} \times \frac{1}{a} = \frac{1}{2a} + \frac{1}{6a}$ if 'a' is odd, and $\frac{1}{3} \times \frac{1}{a} = \frac{1}{2}\left(\frac{2}{3} \times \frac{1}{a}\right) = \frac{1}{4a} + \frac{1}{12a}$ and

if 'a' is even and equals 2c, then $\frac{2}{3} \times \frac{1}{2c} = \frac{1}{3c}$,

and $\frac{1}{3} \times \frac{1}{2c} = \frac{1}{6c}$.

3) If $ab = c$, then $\frac{1}{c} \times b = \frac{1}{a}$.

4) $\frac{1}{a}\left(\frac{1}{c} + \frac{1}{d}\right) = \frac{1}{ac} + \frac{1}{ad}$. The Distributive Property of Multiplication over Addition.

5) $\frac{1}{a} \times \frac{1}{b} = \frac{1}{b} \times \frac{1}{a}$. The Commutative Property of Multiplication.

6) $1 \times \frac{1}{b} = \frac{1}{b}$. The Identity of Multiplication.

When the Egyptian multiplied two sets of fractions, he used the

previous rules and the properties above to find the answer. He multiplied

each fraction from the first set by all the fractions of the second set; then he

added all the results to be the answer.

Problem 32 (RMP)

This is a good example for applying some of the previous properties and rules. In this problem, the scribe proved his result by showing that

$$\left(1 + \tfrac{1}{6} + \tfrac{1}{12} + \tfrac{1}{114} + \tfrac{1}{228}\right)\left(1 + \tfrac{1}{3} + \tfrac{1}{4}\right) = 2.$$

The scribe used the Commutative Property of Multiplication to multiply $\left(1 + \tfrac{1}{3} + \tfrac{1}{4}\right)$ by $\left(1 + \tfrac{1}{6} + \tfrac{1}{12} + \tfrac{1}{114} + \tfrac{1}{228}\right)$. Then he used the Distributive Property of Multiplication over Addition as follows:

$1 \rightarrow \boxed{1} \; \overline{6} \; \tfrac{1}{12} \; \tfrac{1}{114} \; \tfrac{1}{228}$	Identity of Multiplication	
$\tfrac{2}{3} \rightarrow \tfrac{2}{3}\,\tfrac{1}{9}\,\tfrac{1}{18}\,\tfrac{1}{171}\,\tfrac{1}{342}$	Rule 2	
$\tfrac{1}{3} \rightarrow \boxed{\tfrac{1}{3}} \; \tfrac{1}{18}\,\tfrac{1}{36}\,\tfrac{1}{342}\,\tfrac{1}{684}$	Rule 1, 2	
$\tfrac{1}{2} \rightarrow \tfrac{1}{2}\,\tfrac{1}{12}\,\tfrac{1}{24}\,\tfrac{1}{228}\,\tfrac{1}{456}$		
$\tfrac{1}{4} \rightarrow \boxed{\tfrac{1}{4}} \; \tfrac{1}{24}\,\tfrac{1}{48}\,\tfrac{1}{456}\,\tfrac{1}{912}$		

(1) The total is $1\tfrac{1}{6}\tfrac{1}{3}\tfrac{1}{4} = 1 + \tfrac{1}{2} + \tfrac{1}{4}$ from the rule of addition, and the (2) remainder is the series of smaller fractions which are:

$$\tfrac{1}{12} + \tfrac{1}{114} + \tfrac{1}{228} + \tfrac{1}{18} + \tfrac{1}{36} + \tfrac{1}{342} + \tfrac{1}{684} + \tfrac{1}{24} + \tfrac{1}{48} + \tfrac{1}{456} + \tfrac{1}{912}.$$

(3) Apply 912 as a common denominator:

$$76 \quad 8 \quad 4 \quad 50\tfrac{2}{3}\;\; 25\tfrac{1}{3}\;\; 2\tfrac{2}{3}\;\; 1\tfrac{1}{3}\;\; 38 \quad 19 \quad 2 \quad 1 = 228.$$

Find 228 of 912:

$$1 \longrightarrow 912$$
$$\tfrac{1}{2} \longrightarrow 456$$
$$\tfrac{1}{4} \longrightarrow 228$$

$$228 \text{ of } 912 = \tfrac{1}{4}$$

$$\text{The total: } 1\tfrac{1}{2}\tfrac{1}{4}\tfrac{1}{4} = 2$$

The 3 steps of multiplication are:

1. Select the product that would make the sum less than the required product, but nearly equal to it.
2. Find the remainder.
3. Find the method to find the remainder.

> This method is called the second kind of multiplication

DIVISION OF FRACTIONS

The process of division of fractions was very close to the process of multiplication of fractions. If we want to divide 'a' by $\frac{1}{b}$, then by what must we multiply $\frac{1}{b}$ to get 'a'? In the same problem, when Ahmes wanted to divide 2 by $1\frac{1}{3}\frac{1}{4}$, he tried to find the number that would be multiplied by $1\frac{1}{3}\frac{1}{4}$ to get 2. He proceeded as follows:

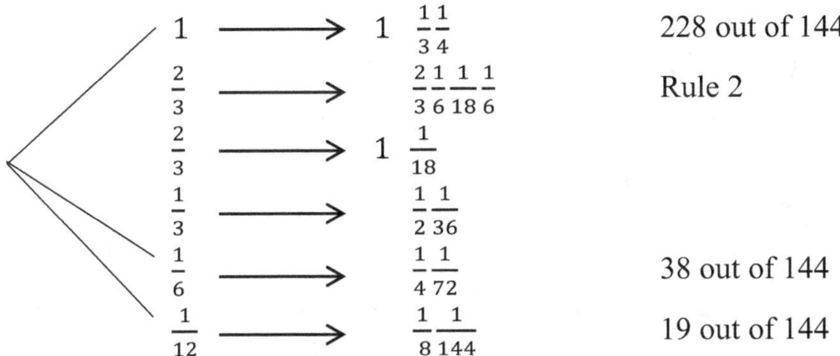

The largest denominator of the smallest fraction is 144. The sum of the numbers that correspond to the multipliers checked is

$228 + 38 + 19 = 285$ out of 144. To get 2, the scribe needed 288 out of 144.

The scribe needed to take $288 - 285 = 3$ out of 144. To get that, he multiplied.

$$1 \longrightarrow 144$$
$$\tfrac{1}{3} \longrightarrow 48$$
$$\tfrac{1}{4} \longrightarrow 36$$

The total is 228.

$$228 = \left(1 + \tfrac{1}{3} + \tfrac{1}{4}\right)144 \qquad \text{Using Rule 3 if } ab = c, \text{ then } \tfrac{1}{c}b = \tfrac{1}{a}$$
$$\tfrac{1}{144}\left(1 + \tfrac{1}{3} + \tfrac{1}{4}\right) = \tfrac{1}{228}$$

To continue our first multiplication:

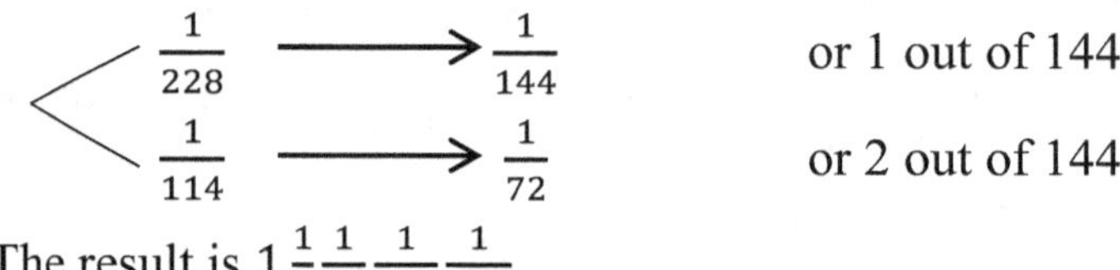

The result is $1\tfrac{1}{6}\tfrac{1}{12}\tfrac{1}{114}\tfrac{1}{228}$

The Solution's Summary

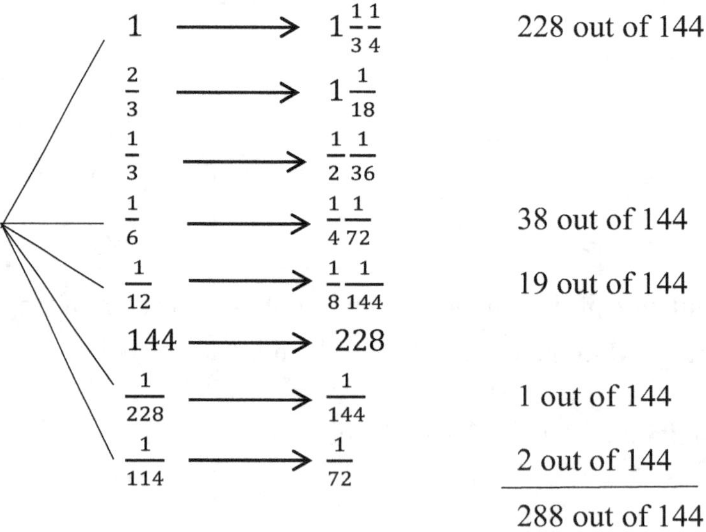

288 out of 144

The scribe used the third rule: If $ab = c$, then $\frac{1}{c} \times b = \frac{1}{a}$. By using this rule, the scribe can use any unit fraction needed as a multiplier in the solution of the problems, other than $\frac{1}{2}$ and its power or $\frac{2}{3}$.

The fifth operation is multiplying a number by itself some given number of times: a^n. The ancient Egyptian introduced to us the geometric series, 2^n. They showed us that any positive integer can be expressed uniquely as the sum of some terms of the series. Then they developed their multiplication procedure by continued doubling, assuming any multiplier can be expressed uniquely as the sum of some terms of 2^n. In modern times, this series is very important in developing the binary system and developing the arithmetic operations performed by modern computers. The ancient Egyptians also introduced to us, in their papyri, other geometric progressions such as in **Problem 97 (RMP),** where the common ratio is 7, and the first term is 7; this geometric progression being 7, 49, 343, 2401, 16,807, and 117,649. Also, **Problem 80 (RMP)** was to express the fractions of hekat in hin: 1, $\frac{1}{2}$, $\frac{1}{4}$, $\frac{1}{16}$, $\frac{1}{32}$, $\frac{1}{64}$. In geometry, the Egyptian found the area of the square by squaring the side of the square. Also, the area of the circle $=$ $\left(\frac{8}{9} \text{ diameter}\right)\left(\frac{8}{9} \text{ diameter}\right) = \left(\frac{8}{9} \text{ diameter}\right)^2$. **Problem 44 (RMP)** was to find the volume of a rectangular granary with a length of 10 cubits, a width of 10 cubits, and a height of 10 cubits. The scribe multiplied the length times width times height $= 10 \times 10 \times 10 = (10)^3 = 1000$ cubic cubits.

The sixth operation is to find some number multiplied by itself (squared) a given number of times to equal some given number. This means getting the 'n' root of any number, $\sqrt[n]{a}$, such that when $n = 2$, we write it $\sqrt[2]{a} = \sqrt{a}$, and we call it the square root of 'a'. There are some problems in the Egyptian papyri where the scribe found the square root of a number to solve these problems.

In the Hieratic Papyri, the square root of the number is always stated, but not calculated. In LV,4 of the Kahun Papyrus, the square root of 16, $\sqrt{16}$, is stated as 4. In the Berlin Papyrus, the square root of $1\frac{1}{2}\frac{1}{16}$ is stated as $1\frac{1}{4}$ and the square root of $6\frac{1}{4}$ is stated as $2\frac{1}{2}$.

In the Demotic Papyri, we found the rules applied in details which were used to find the square roots. These rules are the following:

1) $\sqrt{a^2} = a$

2) If $n = a^2 + b$, then $\sqrt{n} = a + \frac{b}{2a}$; if $n = a^2 - b$, then $\sqrt{n} = a - \frac{b}{2a}$

3) $\sqrt{\frac{a}{b}} = \frac{\sqrt{a}}{\sqrt{b}} = \frac{\sqrt{ab}}{b}$

4) $\sqrt{\frac{a}{b}} = \sqrt{\frac{ac}{bc}} = \frac{\sqrt{ac}}{\sqrt{d^2}} = \frac{\sqrt{ac}}{d}$ if $bc = d^2$

In LV,4 of the Kahun Papyrus, $\sqrt{16} = 4$. The scribe probably found it from a square root table which was similar to the one suggested by the author or the table of squares, and applied Rule 1, $\sqrt{a^2} = a$.

To apply the second rule, for $\sqrt{n}$ the scribe found a and b, such that $n = a^2 + b$. He would then apply the second rule: $\sqrt{a^2 + b} = a + \frac{b}{2a}$.

Problem 10 (BMP 10520)

The scribe found $\sqrt{10} = 3\frac{1}{6}$. By applying Rule 2, $\sqrt{10} = \sqrt{9 + 1}$

$= 3 + \frac{1}{2 \times 3} = 3\frac{1}{6}$.

Problem 9 (Berlin Papyrus)

The scribe found $\sqrt{6\frac{1}{4}} = 2\frac{1}{2}$.

By applying Rule 1 and Rule 3, $6\frac{1}{4}$ is 25 of 4; $\sqrt{25} = 5$, and $\sqrt{4} = 2$.

Then $\sqrt{6\frac{1}{4}} = 5$ from 2, which is 5 out of $2 = 2\frac{1}{2}$.

In the Berlin Papyrus the scribe stated that $\sqrt{1\frac{1}{2}\frac{1}{16}} = 1\frac{1}{4}$; by applying

Rule 1 and Rule 3, $1\frac{1}{2}\frac{1}{16}$ is 25 out of 16, and $\sqrt{25} = 5$ and $\sqrt{16} = 4$.

The result is 5 out of $4 = 1\frac{1}{4}$.

In the Demotic Papyrus, the scribe found $\sqrt{\frac{1}{2}} = \frac{2}{3}\frac{1}{24}$ to make the

denominator a complete square. The scribe used Rule 3 and Rule 2 as follows:

He chose $c = 18$; then $\frac{1}{2} = \frac{18}{36} \Rightarrow \sqrt{\frac{1}{2}} = \frac{\sqrt{18}}{\sqrt{36}} = \frac{\sqrt{18}}{6}$ and made the

denominator a complete square.

Then he used Rule 2 to get $\sqrt{18} = \sqrt{16 + 2} = 4 + \frac{2}{8} = 4\frac{1}{4}$. He then

divided $4 + \frac{1}{2}$ by 6, resulting in $\frac{2}{3}\frac{1}{24}$.

Problem 7 (Cairo Papyrus)

The scribe found $\sqrt{1500} = 38\frac{2}{3}\frac{1}{20}$.

If he used $\sqrt{a^2 + b} = \sqrt{1444 + 56} = 38 + \frac{56}{76} \approx 38\frac{2}{3}\frac{1}{14}$.

If he then uses $\sqrt{a^2 - b} = \sqrt{1521 - 21} = 39 - \frac{21}{78} \approx 38 + \frac{57}{78} \approx 38\frac{2}{3}\frac{1}{20}\frac{1}{70}$.

The scribe rounded the amount to $38\frac{2}{3}\frac{1}{20}$.

The Egyptian Arithmetic System is a decimal system containing the natural numbers and fraction numbers. They define the six operations on this system. This system, with its operations, is the basis for elementary arithmetic now. The methods and procedures to perform these operations may seem to be strange or troublesome to use in our modern times, but within this system, the methods and procedures were based on sound mathematical properties and theorems, as has been shown. The author also shows that the methods they applied have more effect in mathematical reasoning than they did in the methods of arithmetic. The methods of performing the operations were used in the Old World for many thousands of years, and some of these methods are used in our modern times. The methods of multiplication were widely adopted by the Greeks, with some modification, and the usage continued into the Middle Ages in Europe.

When our system of Arabic numerals was perfected, the rules of present day arithmetic were improved. We also developed a complete decimal system with its numbers. Within this new system, the methods of performing the operations were improved. But the basics of the elementary arithmetic, with some modification, are still the same as was found written by the ancient Egyptian.

EGYPTIAN MULTIPLICATION AND MODERN COMPUTERS

Binary Numbers (System)

A decimal system is a system of notations to the base 10 expressed with 0, 1, 2, 3, 4, 5, 6, 7, 8, 9 where each place corresponds to the power of 10.

$$7502 = 7 \times 1000 + 5 \times 100 + 0 \times 10 + 2 \times 1.$$
$$= 7 \times 10^3 + 5 \times 10^2 + 0 \times 10^1 + 2 \times 10^0.$$

We know the ancient Egyptian, in their multiplication, repeatedly multiplied the multiplier by 2^n (doubling) and adding up the intermediate multipliers until they reached the original multipliers and use:

1) The series of 2^n for $n = 0, 1, 2, \ldots$.

2) The theorem that every integer can be expressed as the sum of some terms of 2^n for some 'n'.

The ancient Egyptians, in their multiplication, put the number forth and doubled: ...

... 32	16	8	4	2	1
... 1×2^5	1×2^4	1×2^3	1×2^2	1×2^1	1×2^0

Example:

$$19_{10} = 1 \times 2^4 + 0 \times 2^3 + 0 \times 2^2 + 1 \times 2^1 + 1 \times 2^0 = 16 + 0 + 0 + 2 + 1$$
$$= \quad (1 \quad\quad 0 \quad\quad 0 \quad\quad 1 \quad\quad 1)_2 \quad \text{in binary base 2.}$$

With this method we get the binary representation for any number with 2 symbols (0, 1) and the series of 2^n where $n = 0, 1, 2, 3, \ldots$.

A binary system is a numerical notation to the base 2 where each number, expressed 0 and 1, with a position corresponding to the power of 2. Another example:

$$21_{10} = 1 \times 2^4 + 0 \times 2^3 + 1 \times 2^2 + 0 \times 2^1 + 1 \times 2^0 = 16 + 0 + 4 + 0 + 1$$
$$= \quad (1 \quad\quad 0 \quad\quad 1 \quad\quad 0 \quad\quad 1)_2 \quad \text{in binary base 2.}$$

We can transfer any decimal number to a binary number; we can also transfer any binary number to a decimal number.

$$(1\,0\,1\,1\,1)_2 = 1 \times 2^4 + 0 \times 2^3 + 1 \times 2^2 + 1 \times 2^1 + 1 \times 2^0$$
$$= \quad 16 \quad + \quad 0 \quad + \quad 4 \quad + \quad 2 \quad + \quad 1 = 23_{10}$$

Decimal: 0   1   2   3   4   5   6   7   8   16   32
Binary: 0   1   10 11 100 101 110 111 1000 10000 100000

Operation (addition) on binary number.

In decimals, we add the number and every 10 we add 1 to the next higher number. In binary, we add numbers and every 2, we add 1 to the next higher number.

$$0 + 0 = 0 \quad 1 + 0 = 1 \quad 1+1 = 10 \quad 1+1+1 = 11 \quad 1+1+1+1 = 100 \text{ in binary.}$$

Example: addition

```
  1 *1*                    1* 1*              * = number carried
  0 1 1 0 1 = 13             1 0 1 0 1   =  21
  0 1 1 1 0 = 14             1 0 1 0 1 0 =  42
  1 1 0 1 1 = 27         1 0 1 0 1 0 0 0 0 = 336
                        (1 1 0 0 0 1 1 1 1)₂ = 399
```

Multiplication on binary number

Example: multiply $(1101)_2 \times (1011)_2$

Multiplication of Binary   **Egyptian Multiplication Binary**

```
              1 1 0 1                         1            1 1 0 1
              1 0 1 1             1 0 1 1      10        1 1 0 1 0       1 0 0 0 1 1 1 1
Carried  1 1 1 1 1 0 1                        100       1 1 0 1 0 0
              1 0 1                           1000    1 1 0 1 0 0 0
            0 0 0 0
            1 1 0 1              Decimal:  1    2   2²    2³    2⁴ ... doubling
      1 0 0 0 1 1 1 1           Binary:    1   10  100  1000  10000 ... doubling
```

A computer is composed of electrical circuits, or switches, which are capable of only two states: they are either open (1) or closed (0).

When a decimal number (19) is put into the computer, it will change to a binary number $(1\,0\,0\,1\,1)_2$. This will activate switches.

This means that (1) in the first and (1) in the second positions show the switches are open; (0) in the third and fourth positions show the switches are closed; and (1) in the fifth position shows the switch is open. Any number can be represented by a series of switches either open or closed.

To show that the ancient Egyptian method of multiplication and division is identical to the binary system used in computers, we will multiply 19 by 21. $(10011)_2$ by $(10101)_2$.

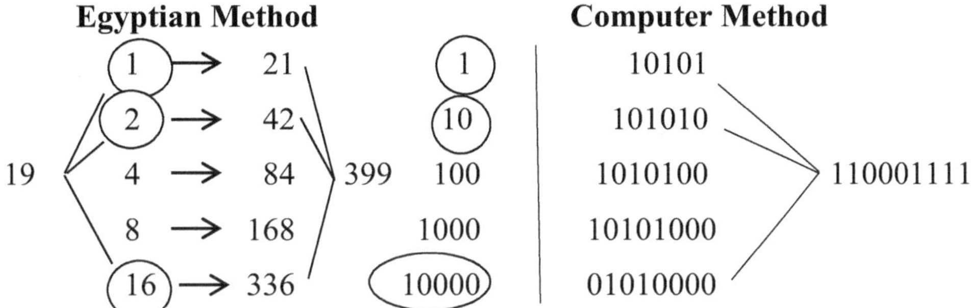

In the computer, you will find the answer in binary (110001111); the computer transfers it to decimal.

$$1 \times 2^0 + 1 \times 2^1 + 1 \times 2^2 + 1 \times 2^3 + 0 \times 2^4 + 0 \times 2^5 + 0 \times 2^6 + 1 \times 2^7 + 1 \times 2^8$$
$$1 \quad + \quad 2 \quad + \quad 4 \quad + \quad 8 \quad + \quad 0 \quad + \quad 0 \quad + \quad 0 \quad + 128 + 256 = 399$$

If the Egyptian wanted to divide 399 by 19, then what is the number multiplied by 19 to get 399? This is the same Egyptian method as multiplying by 19 to get 399.

Egyptian Method **Computer Method**

The answer, $(10101)_2, = 1 \times 2^0 + 0 \times 2^1 + 1 \times 2^2 + 0 \times 2^3 + 1 \times 2^4 = 21.$

Dr. Maher Y. Shawer

The ancient Egyptian methods of multiplication and division, also used by the computer, are identical; they did not need any multiplication tables, no long division, and no borrowing of numbers.

Chapter 8

The Algebra of Egypt

Algebra is the generalization of arithmetic. Algebraic problems are classified as any problems which are solved by the algebraic process, even though it was first solved by merely guessing or by some cumbersome arithmetic process. The science of algebra was known to the ancient Egyptian about 1800 BC or earlier.

The first record we have of algebra, as defined above, is in the Mosco and Ahmes Papyri. Other Hieratic and Demotic Papyri problems can also be classified as algebraic problems. Many of the solutions of these problems required the knowledge of algebraic identities and the solution with algebraic equations. We also found the Egyptians developed formulas for the summations of the special arithmetic and geometric progressions.

In the Pefsu problems, they applied the ratio, proportion, and the rule of three in the solution of the problems. In quantity problems, they used the general solutions for the first and the second degree equations and, also, the general solution of simultaneous equations. They referred to the unknown by "aha" or "heap", which translated to quantity. In solving these equations, two methods were introduced: the method of applying the general rules and the method of false position in solving these equations. The **general rules** they applied show a sophisticated level, even by our standards now. The author will show that these rules are mathematically sound by showing the proofs of the rules. **A <u>false position method</u> is to take a specific value, most likely a false one, assumed for 'x', and the operations indicated on the left-hand side of the equality sign are performed on this assumed number. The result of these operations is then compared with the result desired, and by the use of proportions, the correct answer is found. This method is not found in modern textbooks.**

To get the terms of the <u>arithmetic progression</u> if the sum of the progression s_n is known and the common difference, d, is known, the

Egyptian scribe found the average which is equal to $\frac{s_n}{n}$; then he got the largest term $\frac{s_n}{n} + (n-1)\frac{d}{2}$ by adding one half of the common difference as many times as there were differences. Once he obtained the largest term, he got the others by subtracting the common difference a sufficient number of times as was needed, and the terms in descending order. The scribe applied the rule that the sum of any number of terms, s_n, in any geometric progression, where the first term is equal to the ratio is:

$$s_n = r\left(\frac{L-1}{r-1}\right), \quad s_n = r(s_{n-1}+1) \quad \Rightarrow \quad s_{n-1}+1 = \frac{L-1}{r-1}$$

($L = n$ term; $s_n = $ the sum of $'n'$ terms)

In the arithmetic progression in the Demotic Papyrus, the scribe applied the following:

1) If $s_n = 1 + 2 + 3 + \cdots + n = \displaystyle\sum_{i=1}^{n} i = \frac{n^2+n}{2}$

2) If $s_n = \displaystyle\sum_{a_1}^{a_n} \sum_{i=1}^{n} i$, then $s_n = \left(\frac{n+2}{3}\right)\left(\frac{n^2+n}{2}\right)$

The sequence is a_1 , a_2 , a_3 , ... a_n , such that $a_n = \displaystyle\sum_{i=1}^{n} i$.

The sequence numbers are 1, 3, 6, 10, 15, 21, 28, 36, 45, 55, ... , which are called triangular numbers. The sum of this series is

$$1 + 3 + 6 + 10 + 15 + 21 + \ldots - \sum_{i=1}^{n} i$$

$$s_n = \sum_{a_1}^{a_n} \sum_{i=1}^{n} i = \left(\frac{n+2}{3}\right)\left(\frac{n^2+n}{2}\right) = \frac{n(n+1)}{2}\left(\frac{n+2}{3}\right) = \left(\frac{n+2}{3}\right)\sum_{i=1}^{n} i$$

THE PEFSU

The concepts of ratio and proportion are important in proving many theories in mathematics and in other fields. "The Greek writer, Nicomachos, included ratio in his arithmetic, Eudoxas in geometry, and Theon of Smyrna in his chapter in music and now ratio and proportions, considered topics in algebra" (Smith, 1958, p. 477).

In **Problem 69** in the Ahmes Papyrus, the concept of pefsu was introduced regarding the proportion of the number of loaves of bread or the number of jugs of beer to the number of hekats of grain.

However, the word pefsu is derived from the stem of the verb, to cook. It means something like 'cooking ratio'; that is, the number of units of food or drink that could be made from a unit of material in the problem of cooking, and it determines the relative value of any food or drink (Peet, 1923, p. 22).

Problems 69 to 78 (RMP) are problems about pefsu.

In these problems, the scribe defined the pefsu of bread or beer produced as the following:

$$\text{Pefsu } [p] = \frac{\text{number of loaves of bread or jugs of beer } [n]}{\text{number of hekats of grain needed in the loaves of bread or jugs of beer } [h]}$$

$$p = \frac{n}{h} \Rightarrow n = ph \text{ and } h = \frac{n}{p}$$

This means that the smaller the pefsu of beer, the stronger the beer; and the larger the pefsu, the weaker the beer. This also implies that the smaller the pefsu of the loaf of bread, the heavier the bread; and the larger the pefsu, the lighter the bread.

Problem 69 (RMP)

Three and one half hekats of meal is made into 80 loaves of bread. Determine the amount of meal that is in each loaf and the pefsu.

The scribe solved this problem by first finding the pefsu. Then he found the amount of meal in each loaf. He found the pefsu by applying the formula:

$$\text{Pefsu} = \frac{\text{number of loaves}}{\text{number of hekats}}$$

Multiply by $3\frac{1}{2}$ as to get 80.

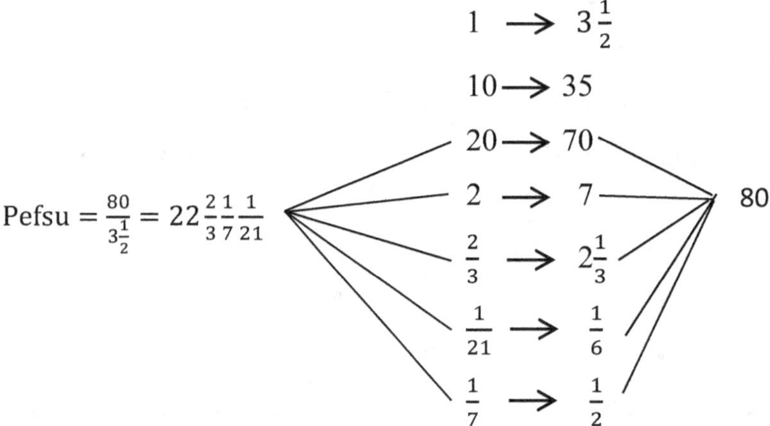

$$\text{Pefsu} = \frac{80}{3\frac{1}{2}} = 22\frac{2}{3}\frac{1}{7}\frac{1}{21}$$

This gives the pefsu of $22\frac{2}{3}\frac{1}{7}\frac{1}{21}$.

Then, the scribe transferred $3\frac{1}{2}$ hekats to ros by multiplying $3\frac{1}{2} \times 320$.

Total:	1120 ro.

To answer the second part of the question, the scribe divided 1120 by 80. To prove that, he then multiplied by 80 to get 1120.

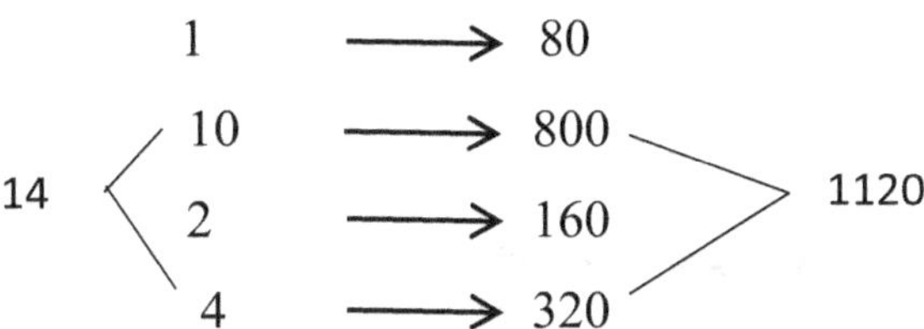

Total: 14

Thus, he determined that one of the loaves of bread contains 14 ro or $\frac{1}{32}$ hekat and 4 ro.

To prove that, he multiplied 80 by $\frac{1}{32}$ hekat and 4 ro to get $3\frac{1}{2}$ hekat.

Problem 71 (RMP)

From 1 des measure of beer, $\frac{1}{4}$ has been poured out. Then, this amount has been filled with water. What is the pefsu of the diluted beer?

(The solution assumes that the amount of grain in 1 des of beer is $\frac{1}{2}$ hekat. The beer was made from besha, which is a kind of grain they used to make the beer.) (Chace, 1927, p. 108.)

The Solution

Take $\frac{1}{4}$ of $\frac{1}{2}$ hekat. The result is $\frac{1}{8}$ hekat. The remainder is $\frac{1}{4}\frac{1}{8}$ hekat.

Multiply $\frac{1}{4}\frac{1}{8}$ to get 1.

The result is $2\frac{2}{3}$.

This is the pefsu which equals $\dfrac{\text{the number of des of beer}}{\text{the number of hekat of the grain left}}$.

Problem 70 (RMP) is similar to Problem 69.

In the next problems, the author will summarize the scribe's solutions to show how he followed the rule that the author will develop and introduce.

Problem 72 (RMP)

Suppose 100 loaves with pefsu 10 are to be exchanged by a number of loaves with pefsu 45. How many of these loaves are there?

The Solution

The scribe used the following:

If $\frac{x}{p_1} = \frac{y}{p_2}$ then $y = \left(\frac{p_2-p_1}{p_1}\right)x + x$

The Proof

$$\frac{x}{p_1} = \frac{y}{p_2} \Rightarrow \frac{y}{x} = \frac{p_2}{p_1} \Rightarrow \left(\frac{y-x}{x}\right) = \frac{p_2-p_1}{p_1}$$

$$y - x = \left(\frac{p_2-p_1}{p_1}\right)x \Rightarrow y = \left(\frac{p_2-p_1}{p_1}\right)x + x$$

Then given 'x' as the number of loaves with p_1 and 'y' as the number of loaves with p_2, we have $x = 100$, $p_1 = 10$, and $p_2 = 45$.

The Solution

Take the excess of 45 over $10 = p_2 - p_1 = 35$

Divide 35 by $10 = \frac{p_2-p_1}{p_1} = \frac{35}{10} = 3\frac{1}{2}$

Multiply 100 by $3\frac{1}{2} = \left(\frac{p_2-p_1}{p_1}\right)x = 3\frac{1}{2} \times 100 = 350$

Add 100 to $350 = \frac{(p_2-p_1)x}{p_1} + x = 450 = y = $ number of loaves

There is an exchange of 100 loaves of pefsu 10 for 450 loaves of pefsu 45.

Problem 73 (RMP)

100 loaves at pefsu 10 are to be exchanged for loaves at pefsu 15. How many loaves will be exchanged?

Exchange the number of loaves 'x' of pefsu 'p' by dividing the grain to make half for a number of loaves n_1 of p_1 and half for n_2 of p_2.

The scribe used the following:

If $\dfrac{x}{p_1} = \dfrac{y}{p_2}$ then $y = \left(\dfrac{p_2}{p_1}\right) x$

x = number of loaves with p_1 and y = number of loaves with p_2.

The rule of three enables one to find one of the numbers that equals two proportions $\left(\dfrac{a}{b} = \dfrac{c}{d}\right)$ if the other three numbers are given.

$x = 100$; $p_1 = 10$; $p_2 = 15$, and y is the number of loaves with $p_2 = 15$.

The scribe applies $\dfrac{x}{p_1} = \dfrac{y}{p_2} \Rightarrow y = \left(\dfrac{x}{p_1}\right) p_2$. (The number of hekat $\times$ p_2.)

The Solution

The amount of flour in these 100 loaves $= \dfrac{100}{10} = \dfrac{x}{p_1} = 10$ hekat.

$y = 10 \times 15 = \left(\dfrac{x}{p_1}\right) p_2 = 150$ loaves.

From the last two problems, the scribe was familiar with the rule of three, and if two ratios are equal then four numbers are proportional, such as $x : p_1 = y : p_2$. Then, various other proportions can be derived from this proportion, such as:

if $\dfrac{x}{p_1} = \dfrac{y}{p_2}$ then $\dfrac{y}{x} = \dfrac{p_2}{p_1}$ and $\dfrac{y-x}{x} = \dfrac{p_2-p_1}{p_1}$.

Problem 74 (RMP)

1000 loaves of pefsu 5 are exchanged, half for loaves of pefsu 10, and half for loaves of pefsu 20. How many of each will there be?

The number of hekats of grain to make n = 1000 loaves with

$p = 5$ is $\dfrac{n}{p} = \dfrac{1000}{5} = 200$ hekats.

Half of the hekats of the grain = 100 hekats.

$h_1 = 100; \quad p_1 = 10; \quad h_2 = 100; \quad p_2 = 20; \quad h = h_1 + h_2$

The number of loaves n_1 from the first half $= h_1 p_1 = 100 \times 10$

$$= 1000 \text{ loaves.}$$

The number of loaves n_2 from the second half $= h_2 p_2 = 100 \times 20$

$$= 2000 \text{ loaves.}$$

The following problem is an exchange of loaves 'n' of pefsu 'p' by an equal number of two types of bread with p_1 and p_2 , and $p_2 = kp_1$.

The scribe found the number of each type by using the following:

the reciprocal of $\left(\dfrac{1}{p_1} + \dfrac{1}{p_2}\right) = \dfrac{p_1 p_2}{p_1 + p_2} = \dfrac{p_1(kp_1)}{p_1 + kp_1} = \dfrac{kp_1{}^2}{p_1(k+1)} = \dfrac{kp_1}{k+1}$

$= \dfrac{1}{2}$ (harmonic mean of p_1, p_2) $= \dfrac{1}{2}\left(\dfrac{2p_1 p_2}{p_1 + p_2}\right) = \dfrac{p_1 p_2}{p_1 + p_2}$

$n = \left(\dfrac{kp_1}{k+1}\right) h.$

Problem 76 (RMP)

1000 loaves of pefsu 10 are to be exchanged for a number of loaves of pefsu 20 and the same number of pefsu 30. How many of each kind will there be?

$$n = 1000; \quad p = 10; \quad p_1 = 20; \quad p_2 = 30 = 1\tfrac{1}{2}p_1 \Rightarrow k = 1\tfrac{1}{2}$$

<u>The Solution</u>

$$\frac{1}{p_1} + \frac{1}{p_2} = \frac{1}{20} + \frac{1}{30} = \frac{2\tfrac{1}{2}}{30} = \frac{1}{12}.$$

The reciprocal $= \dfrac{kp_1}{k+1} = \dfrac{\left(1\tfrac{1}{2}\right)20}{2\tfrac{1}{2}} = 12.$

The number of hekats of the grain with $p = 10$

$$\frac{1000}{10} = 100 \ \text{hekats} \qquad\qquad \frac{n}{p}$$

The number of the loaves of bread of each type

$$= 12h = 12 \times 100 = 1200 \ \text{loaves} \qquad \left(\frac{kp_1}{k+1}\right)n$$

The first type of loaf with pefsu 20 will take

$$\frac{1200}{20} = 60 \ \text{hekats.} \qquad\qquad \left(\frac{kp_1}{k+1}\right)n \div p_1$$

The second type of loaf with $p_2 = 30,$ will take

$$\frac{1200}{30} = 40 \ \text{hekats.} \qquad\qquad \left(\frac{kp_1}{k+1}\right)n \div p_2$$

This problem is interesting because it is solved by adding the reciprocals of given numbers for finding the harmonic mean.

Problem 77 (RMP)

Suppose the problem stated 10 des of beer of 2 pefsu are to be exchanged for a number of loaves of 5 pefsu. How many loaves will there be?

The Solution

10 des of beer which takes 5 hekats of flour can be exchanged for 25 loaves of bread of 5 pefsu.

$\frac{x}{5} = 5 \Rightarrow x = 25$ loaves.

This translates as follows:

The number of hekat $\frac{10}{2} = 5$. Let 'x' be the number of loaves.

$\frac{x}{5} = 5 \Rightarrow x = 25$ loaves.

Problem 63 (RMP)

Divide 700 loaves among four men in a proportion with the numbers $\frac{2}{3}$, $\frac{1}{2}$, $\frac{1}{3}$, and $\frac{1}{4}$.

Let me know the share each man receives.

The method to solve this problem is when the scribe finds the sum of the given proportions are greater than 1.

The Solution

The scribe added the proportions he finds to be greater than 1. He gets 1 by operating on the sum of the proportions; then he reduced 'n' to 'n_1'; then he found the proportion from 'n_1' as follows:

$$n = 700 \qquad \frac{2}{3} + \frac{1}{2} + \frac{1}{3} + \frac{1}{4} = 1\frac{11}{24}.$$ The sum of the proportion > 1.

Divide 1 by $1\frac{1}{2}\frac{1}{4}$; the result is $\frac{1}{2}\frac{1}{14}$.

$\frac{1}{2}\frac{1}{14}$ of $700 = 400 = n_1$

$\frac{2}{3}$ of $400 = 266\frac{2}{3}$ $\qquad$ $p_1 \times n_1$ $\quad$ loaves

$\frac{1}{2}$ of $400 = 200$ $\qquad$ $p_2 \times n_1$ $\quad$ loaves

$\frac{1}{3}$ of $400 = 133\frac{1}{3}$ $\qquad$ $p_3 \times n_1$ $\quad$ loaves

$\frac{1}{4}$ of $400 = 100$ $\qquad$ $p_4 \times n_1$ $\quad$ loaves

Total $= 700$ loaves.

Ten of the twenty-five problems of the Mosco Papyrus are pefsu problems.

Solving Equations

A <u>false position method</u> is to take a specific value, most likely a false one, assumed for 'x', and the operations indicated on the left-hand side of the equality sign are performed on this assumed number. The result of these operations is then compared with the result desired, and by the use of proportions, the correct answer is found.

THE FIRST DEGREE EQUATION

If $ax = m$.

The ancient Egyptians solved the first degree equation by two methods; the Rule of Single False Position and the formula for solving 'x'.

THE RULE OF SINGLE FALSE POSITION

For $ax = m$

We assume the false value for $x = t$

Then $at = n$

$$\frac{ax}{at} = \frac{m}{n}$$

$$\frac{x}{t} = \frac{m}{n}$$

$$x = \left(\frac{m}{n}\right)t = \left(\frac{t}{n}\right)m$$

For example, to the equation $4x = 12$.

We assume that $x =$ (test value) or (false value) $= t = 2$

We replace 'x' by 2 on the left side of the equation.

We call 2 the false value or the test value.

By replacing $x = 2$, then $4x = 8$; $m = 12$, $n = 8$, $t = 2$

Since $\dfrac{m}{n} = \dfrac{12}{8} = \dfrac{3}{2}$, the real value $= \dfrac{m}{n}$ (t).

The real value $= \dfrac{3}{2}$ (2) $= 3$.

3 will be the solution for 'x' in the equation $4x = 12$.

The false position method was used by the ancient Egyptians in the Ahmes Papyrus (1685-1650 BC) in problems 24, 25, 26, 27, 28, 29, and 30.

The next recorded document using the false position method was an Indian mathematical text by Vaishali Ganit (c. 3rd century BC) and then a Chinese text called the "Nine Chapters of Mathematical Art" dated from 200 BC to 100 AD. It was used by Arab mathematicians such as Abu Kamil (c. 900 AD), Sinan Ibn-al-Fath (10th century), Albanna (c. 1300 AD), and various others. The Arabs called the method the Hisab El-Catayan. As noted by Smith (1958), Suma Paciolis, in 1494 named this method El-Catayan. European writers of the 16th Century used the Paciolis name. After the 16th Century, the name of El-Catayan disappeared.

In general, then, the method was called by such names as "Rule of False," "Rule of Position," and "Rule of False Position." It is generally called the "Method of False Position" (Smith, 1958, p. 437).

The following problems were taken from ancient Egyptian papyri.

Problem 24 (RMP)

A quantity and $\frac{1}{7}$ of that quantity, added together, became 19. What is the quantity?

This translated to the following equation: $x + \frac{1}{7}x = 19$, what is x?

<u>The Solution</u>

Assume the quantity is 7.   t = 7 (test value), m = 19

$$1 \longrightarrow 7$$
$$\frac{1}{7} \longrightarrow 1$$

Total   8 = n.

How many times must 8 be multiplied to give 19?

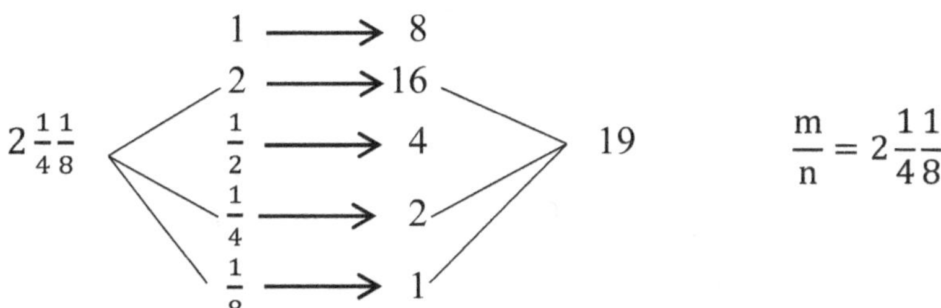

8 must be multiplied by $2\frac{11}{48}$ to give 19; therefore, the answer is $2\frac{11}{48}$.

Then we have to multiply 7 by $2\frac{11}{48}$ to find the quantity.

$$7 \begin{cases} 1 \longrightarrow 2\frac{11}{48} \\ 2 \longrightarrow 4\frac{11}{24} \\ 4 \longrightarrow 9\frac{1}{2} \end{cases} \qquad x = \frac{t}{n}m = \left(2\frac{11}{48}\right)(7) = 16\frac{11}{28}$$

$$16\frac{11}{28} \quad \text{is the answer.}$$

The Proof

$$1 \rightarrow 16\frac{1}{2}\frac{1}{8}$$

$$\frac{1}{7} \rightarrow 2\frac{1}{4}\frac{1}{8}$$

The total is 19

Problem 29 from RMP

If a quantity and $\frac{2}{3}$ of that quantity are added together and $\frac{1}{3}$ of the sum is added, and $\frac{1}{3}$ of the new sum is added with a result of 10, what is the quantity?

$$\frac{1}{3}\left[\left(x+\frac{2}{3}x\right)+\frac{1}{3}\left(x+\frac{2}{3}x\right)\right]=10=m.$$

<u>The Solution</u>

Let 't' $= 27$; then the result $= 20 = n$.

As many times as 20 must be multiplied to give 27.

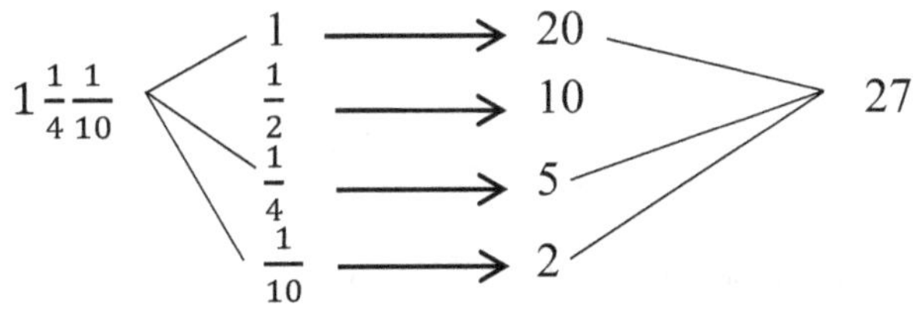

$$\frac{t}{n}=1\frac{1}{4}\frac{1}{10}=\frac{27}{20}$$

20 has to be multiplied by $1\frac{1}{4}\frac{1}{10}$ to give 27.

10 must also be multiplied by $1\frac{1}{4}\frac{1}{10}$ to give the required number.

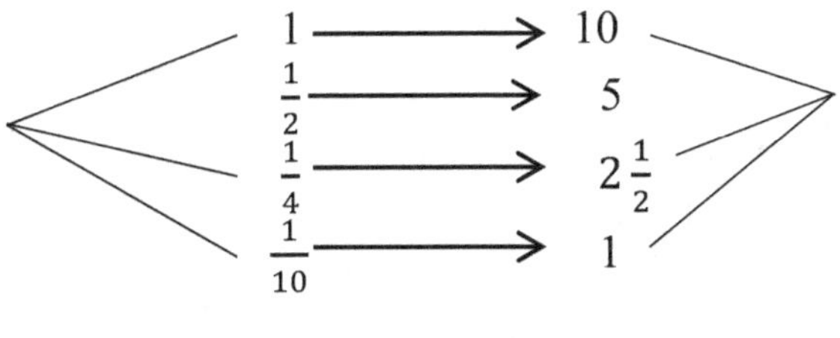

$$x=\frac{t}{n}m=\left(1\frac{1}{4}\frac{1}{10}\right)(10)=13\frac{1}{2}$$

The required number is $10+2\frac{1}{2}+1=13\frac{1}{2}$.

After most of the problems were solved, the scribe checked the solution and called the checking the proof.

The proof

$$1 \longrightarrow 13\tfrac{1}{2}$$
$$\tfrac{2}{3} \longrightarrow 9$$

$$\text{Total} \qquad 22\tfrac{1}{2}$$
$$\tfrac{1}{3} \longrightarrow 7\tfrac{1}{2}$$

$$\text{Total} \qquad 30$$
$$\tfrac{2}{3} \longrightarrow 20$$
$$\tfrac{1}{3} \longrightarrow 10$$

The last two parts appear in his solution, and he assumes he solved the first part first.

Problem 19 (MMP)

This problem was not solved by the Method of False Position.

The scribe followed the rule:

If $ax + b = c$

Then $x = \tfrac{1}{a}(c-b);$ and if $b = 0,$ then $x = \tfrac{c}{a}.$

Problem 19 for calculating the heaps:

If $1\tfrac{1}{2}$ times the heap, together with 4, becomes 10, what is the heap?

The equation is $1\tfrac{1}{2}x + 4 = 10.$

The Solution

$a = 1\tfrac{1}{2} \qquad b = 4 \qquad c = 10$

Compute the excesses of these 10 over 4, and the result $= 6 \qquad (c-b)$

Calculate $1\tfrac{1}{2}$ until you find 1; the result is $\tfrac{2}{3} \qquad \left(\tfrac{1}{a}\right)$

Then multiply $\tfrac{2}{3}$ by 6; the result $= 4$ heaps. $\qquad \tfrac{1}{a}(c-b)$

FALSE POSITION AND SIMULTANEOUS EQUATIONS

The Berlin Papyrus has two problems which Schack-Schockenburg restored and translated.

The two problems are expressed as the following equation:

1) $x^2 + y^2 = 100$ and $4x - 3y = 0$

2) $x^2 + y^2 = 400$ and $4x - 3y = 0$

In this case, the scribe probably used:

if $x^2 + y^2 = a^2$ and $x = ky$; then $y = \dfrac{a}{\sqrt{k^2+1}}$ and $x = \dfrac{ka}{\sqrt{k^2+1}}$

The Proof

$x^2 + y^2 = a^2, \quad x = ky$

$(ky)^2 + y^2 = a^2 \Rightarrow y^2(k^2 + 1) = a^2$

$y^2 = \dfrac{a^2}{k^2+1} \Rightarrow y = \dfrac{a}{\sqrt{k^2+1}}$

$x = \dfrac{ka}{\sqrt{k^2+1}}$

The first problem is stated as follows:

It is said that the area of a square is 100 square cubits, and this area is equal to that of two smaller squares with one side $\dfrac{1}{2}\dfrac{1}{4}$ of the other side. What are the sides of the other two squares?

$x^2 + y^2 = 100$

$x = \dfrac{3}{4}y \longrightarrow 4x = 3y \quad \text{or} \quad 4x - 3y = 0$

$a^2 = 100 \text{ and } k = \dfrac{3}{4}.$

The scribe solved it as follows:

Assume quantity $y = 1;$ then $x = \left(\frac{1}{2}\frac{1}{4}\right)$

$$x^2 + y^2 = \left(\frac{1}{2}\frac{1}{4}\right)^2 + 1^2 = \frac{1}{2}\frac{1}{16} + 1$$

Total $1\frac{1}{2}\frac{1}{16}$ or $k^2 + 1$

The square root of $1\frac{1}{2}\frac{1}{16} = 1\frac{1}{4}$ $\qquad\qquad \sqrt{k^2 + 1}$

What number multiplied by $1\frac{1}{4}$ makes 10?

$$1 \longrightarrow 1\frac{1}{4} \qquad y = \frac{a}{\sqrt{k^2+1}} = \frac{10}{1\frac{1}{4}} = 8$$

$$2 \longrightarrow 2\frac{1}{2}$$

$$4 \longrightarrow 5$$

$$8 \longrightarrow 10$$

Then one side is $y = 8$

The other side

$$1 \longrightarrow 8 \qquad x = \frac{ka}{\sqrt{k^2+1}} = \left(\frac{3}{4}\right)8 = 6$$

$$\frac{1}{2} \longrightarrow 4$$

$$\frac{1}{4} \longrightarrow 2$$

$x = \left(\frac{1}{2}\frac{1}{4}\right)y \qquad x = 6.$

SOLVING SIMULTANEOUS EQUATIONS USING THE GENERAL RULES

$xy = a \dots (1)$

$x = k\,y \dots (2)$

To solve these equations, you need to know how to solve the first degree equation and need to know how to find the square root. We know that the ancient Egyptians knew these concepts.

To solve this equation, use the value of $y = x\left(\frac{1}{k}\right)$

Then you replace 'y' with $x\left(\frac{1}{k}\right)$ then $x\left(\frac{1}{k}\right)x = a \Longrightarrow$

$x^2 = ak \qquad x = \sqrt{ak} \qquad \therefore \quad y = \frac{x}{k} = \frac{\sqrt{ak}}{k}$

The Egyptian rule will be if two quantities multiplied together equal a number 'a' and the first quantity equals another number multiplied by the second quantity 'k', then the first quantity will equal the square root of the two numbers multiplied together $\sqrt{ak}$, and the second quantity equal to the first quantity multiplied by the reciprocal of the second number.

The problem LV,4 of the Kahun Papyrus which translator, F. L. Griffith wrote and R. J. Gillings suggested, could be summarized as:

$xy = 12 = a$

and $\quad y = \frac{3}{4}x \qquad x = \frac{4}{3}y$

<u>The Solution</u>
Make $\frac{1}{2}\frac{1}{4}$ to find 1;

the result is $1\frac{1}{3}$. $\hspace{3cm} k$

Then $12 \times 1\frac{1}{3} = 16$ $\hspace{2.5cm} ka$

The square root of 16 is $4 = x$ $\hspace{1.5cm} \sqrt{ka} = x$

$\frac{1}{2}\frac{1}{4}$ of $4 = \ 3 = \frac{1}{k}(x) = y.$

This translates to:

Find $k = \frac{4}{3}$

The problem $xy = 12 \quad$ and $x = \frac{4}{3} y$

Then find $x = \sqrt{\frac{4}{3}(12)} = \sqrt{16} = 4$

Then find $y = \frac{3}{4} \; x = \left(\frac{3}{4}\right)4 = 3$

The same methods were used to solve Problems 7, 14, 16, and 18 in the Cairo Papyrus.

Problem 7 (Cairo Papyrus)

Given 1000 cloth cubits as the area of a rectangular sail, with a height $1\frac{1}{2}$ times the width, determine the height and the width.

The scribe of the Cairo Papyrus used the same method as the scribe of LV,4 of the Kahun Papyrus.

$$a = xy = 1000 \text{ cloth cubits} \qquad x = \frac{3}{2}y \qquad k = \frac{3}{2}$$

The Solution

Find its half, when $\frac{3}{2} = 1 + \frac{1}{2}$. The ratio is $1:1\frac{1}{2}$. $[ka = 1\frac{1}{2}(1000) = 1500]$

Reduce it to its square root $x = \sqrt{ka} = \sqrt{1500}$. The result is $38\frac{2}{3}\frac{1}{20}$.

We shall say the height of the sail is $38\frac{2}{3}\frac{1}{20}$ cubits. (This is an approximate answer.)

Take $\frac{2}{3} = \left(\frac{1}{k}\right)$ which is $\frac{2}{3}$ of the height.

The result is $25\frac{2}{3}\frac{1}{10}\frac{1}{20}$ cubits, which is the width. (This is an approximate answer.)

Problem 30 (Cairo Papyrus)

The scribe used the general solution for the equation:

$$x^2 = b^2 + (x - a)^2$$

Which is $x = \dfrac{a^2 + b^2}{2a}$

The Proof

$$x^2 = b^2 + (x - a)^2 = b^2 + x^2 - 2ax + a^2$$
$$\therefore 2ax = b^2 + a^2$$

then $x = \dfrac{a^2 + b^2}{2a}$

Problem 30 (CP)

Given 2 as the number of cubits by which an erect pole has been lowered when its foot is moved outward by 6 cubits, determine the original height of the pole.

$$6^2 + (x - 2)^2 = x^2$$

The scribe used the formula

$$x = \frac{a^2 + b^2}{2a}$$

The Solution

Multiply $b^2 = 6 \times 6 = 36;$ $a^2 = 2 \times 2 = 4;$ $a^2 + b^2 = 4 + 36 = 40;$

$$2a = 2 \times 2 = 4$$

$$40 \div 4 = 10 \quad = \frac{a^2 + b^2}{2a}$$

The original height of the pole = 10 cubits.

Problem 34 (Cairo Papyrus)

The scribe took the general solution of the two equations:

$$x^2 + y^2 = d^2 \qquad (1)$$

$$xy = a \qquad (2)$$

Which is $x = \frac{1}{2}\left(\sqrt{d^2 + 2a} + \sqrt{d^2 - 2a}\,\right)$

$$y = \frac{1}{2}\left(\sqrt{d^2 + 2a} - \sqrt{d^2 - 2a}\,\right)$$

<u>The Proof</u>

$$x^2 + y^2 = d^2 \qquad (1)$$

$$xy = a \Rightarrow 2xy = 2a \qquad (2)$$

$$x^2 + y^2 + 2xy = d^2 + 2a \qquad (3)$$

$$(x + y)^2 = d^2 + 2a$$

$$x + y = \sqrt{d^2 + 2a} \qquad (4)$$

and $x^2 + y^2 - 2xy = d^2 - 2a$

$$(x - y)^2 = d^2 - 2a \rightarrow x - y = \sqrt{d^2 - 2a} \qquad (5)$$

From 4 , 5

$$2x = \sqrt{d^2 + 2a} + \sqrt{d^2 - 2a} \qquad \text{add 4 , 5}$$

$$x = \frac{1}{2}\left(\sqrt{d^2 + 2a} + \sqrt{d^2 - 2a}\,\right)$$

$$2y = \sqrt{d^2 + 2a} - \sqrt{d^2 - 2a} \qquad \text{subtract 5 from 4}$$

$$y = \frac{1}{2}\left(\sqrt{d^2 + 2a} - \sqrt{d^2 - 2a}\,\right)$$

Problem 34 (CMP)

Given a rectangle with an area of 60 square cubits and a diagonal of 13, determine the two sides.

$$x^2 + y^2 = 13^2 = 169 \qquad (1)$$

$$a = xy = 60 \text{ square cubits} \qquad (2)$$

$$d = 13 \text{ and } a = 60$$

The Solution

Multiply $13 \times 13 = 169$	$[\,d^2\,]$
Multiply $2 \times 60 = 120$	$[\,2a\,]$
Add $120 + 169 = 289$	$[d^2 + 2a]$
Take the square root of $289 = 17$	$[\sqrt{d^2 + 2a}\,]$
Take the excess of 169 against 120; the result is 49	$[d^2 - 2a]$
The square root of $49 = 7$	$[\sqrt{d^2 - 2a}\,]$
Take 7 from $17 = 10$	$[\sqrt{d^2 + 2a} - \sqrt{d^2 - 2a}\,]$
$y = \frac{1}{2}$ of $10 = 5$ cubits = the width	$\frac{1}{2}[\sqrt{d^2 + 2a} - \sqrt{d^2 - 2a}\,]$
Add $17 + 7 = 24$	$[\sqrt{d^2 + 2a} + \sqrt{d^2 - 2a}\,]$
$x = \frac{1}{2}$ of $24 = 12$ cubits = the length	$\frac{1}{2}[\sqrt{d^2 + 2a} + \sqrt{d^2 - 2a}\,]$

The Proof

The area of a plot of land is $12 \times 5 = 60$ square cubits.

Multiply 12×12; the result is 144. $\qquad x^2$

Multiply 5×5; the result is 25. $\qquad y^2$

Add 144 and 25; the result 169. $\qquad d^2$

The square root of 169 is 13 which is the diagonal, 'd' in cubits.

In **Problems 8-10** in the Cairo Papyrus, the scribe found 'y' by simultaneously applying the solution of the following equations:

If $hw = a$ $\qquad$ (1)

$(h - x)w = a_1$ $\qquad$ (2)

$(h - x)(w + y) = a$ $\quad$ (3)

then $y = \dfrac{a - a_1}{h - x}$ $\qquad$ or $\qquad$ $y = \dfrac{xw}{h - x}$

Problem 8 (CMP)

Given a piece of cloth 7 cubits high by 5 cubits wide, what must be added to the width when 1 cubit is taken from the height and the area remains the same?

$h = $ height

$w = $ width

$x = $ the number of cubits taken from the height

$y = $ the number of cubits added to the width

$h = 7$ cubits $\quad$ $x = 1$ cubit $\quad$ $w = 5$ cubits

The Solution

$a = 7 \times 5 = 35$ square cubits (the area) $\qquad$ $a = hw = 35$

Take one cubit from the height; the result is 6 $\qquad$ $h - x = 6$

The new area is $a_1 = 6 \times 5 = 30$ square cubits $\qquad$ $a_1 = (h - x)w = 30$

35 (old area) $- 30$ (new area) $= 5$ sq. cubits $\qquad$ $a - a_1 = 35 - 30 = 5$

$y = 5 \div 6 = \dfrac{5}{6}$ cubits $\qquad\qquad$ $\dfrac{a-a_1}{h-x} = \dfrac{5}{6} = y$

The new width is $5 + \dfrac{5}{6} = 5\dfrac{5}{6}$ cubits. $\qquad$ $(w + y) = 5 + \dfrac{5}{6} = 5\dfrac{5}{6}$

The Proof

$a = \left(5\dfrac{5}{6}\right) 6 = 35$ square cubits

$(h - x)\left(w + \dfrac{xw}{h-x}\right)$

$= (h - x)\left(\dfrac{hw-wx+wx}{h-x}\right) = \dfrac{wh-wx+wx}{h-x} = \dfrac{(h-x)wh}{h-x} = wh$

If $hw = a$ $\qquad\qquad$ (1)

and $(h + x)(w - y) = a$ $\qquad$ (2)

$h(w - y) = a_1$ $\qquad\qquad$ (3)

then $x = \dfrac{a-a_1}{w-y} = \dfrac{hy}{w-y}$

The new height is $h + \dfrac{hy}{w-y} = \dfrac{hw}{w-y} = \left(\dfrac{w}{w-y}\right) h.$

Problem 11 (CMP)

Given a piece of cloth 6 cubits high by 4 cubits wide, determine what must be added to the height when 1 cubit is taken from the width and the area remains the same.

$a = 6 \times 4 = 24$ sq. cubits, $\qquad a_1 = 6(4 - 1) = 18$ sq. cubits, $\qquad y = 1$

The scribe found that the value of $x = \dfrac{a-a_1}{w-y} = \dfrac{24-18}{4-1} = \dfrac{6}{3} = 2$ cubits

The new height $= 6 + 2 = 8$ cubits.

The Proof

$a = 8 \times 3 = 24$ square cubits.

ARITHMETIC AND GEOMETRIC PROGRESSION

An arithmetic progression is a sequence in which each term (except the first) differs from the previous one by a constant amount, which is the common difference.

The progression takes the form: a, a + d, a + 2d + ... + a + (n-1) d
'a' is the first term and 'd' is the common difference.

The 'n' term $L = a + (n-1)\, d$, which is the larger share

The sum of the first 'n' terms $s_n = na + \frac{1}{2} n\,(n-1)\, d$.

$$s_n = \frac{1}{2} n\, [2a + (n-1)\, d]$$

$$s_n = \frac{n}{2}\, [a + a + (n-1)\, d] = \frac{n}{2}\, [a + L]$$

The average of the terms is $\frac{s_n}{n} = [\frac{a+L}{2}]$.

$$\frac{s_n}{n} + (n-1)\frac{d}{2} = (\frac{a+L}{2}) + (n-1)\frac{d}{2} = \frac{1}{2}[a + L + (n-1)d]$$

$$= \frac{L+a+(n-1)d}{2} = \frac{2L}{2} = L$$

$$\therefore L = \frac{s_n}{n} + (n-1)\frac{d}{2}.$$

To find the terms of the arithmetic progression, find the largest term which is 'L'. Then subtract the common difference to find the rest of the terms.

Note: **The first definite trace that we have of an arithmetic series, as such, is in the Ahmes Papyrus (c.1550 B.C.), where two problems, Problem 40 and Problem 64, are given involving such a sequence. (Smith, 1953, p. 498.)**

Problem 64 (RMP)

If we want to distribute 10 hekats of barley among 10 men so that the difference of each man and his neighbor, in hekats of barley, is $\frac{1}{8}$ different, then what is each man's share?

The scribe used the last formula for the last terms

$$L = \frac{S_n}{n} + (n-1)\frac{d}{2}$$

$$S_n = 10 \quad n = 10 \quad d = \frac{1}{8}$$

He found the average to be $\frac{S_n}{n} = \frac{10}{10} = 1$.

Then he took $\frac{d}{2} = \frac{1}{8} \times \frac{1}{2} = \frac{1}{16}$ and found the value of $\frac{d}{2}(n-1) = \frac{1}{16} \times 9$

$$= \frac{9}{16} = \frac{1}{2}\frac{1}{16}.$$

$$L = \frac{S_n}{n} + (n-1)\frac{d}{2} = 1\frac{1}{2}\frac{1}{16}.$$

$1\frac{1}{2}\frac{1}{16}$ is the larger share of hekats.

Subtract $\frac{1}{8}$ for each man until you reach the smallest share and the ten portions will be:

$$1\frac{1}{2}\frac{1}{16}, \; 1\frac{1}{4}\frac{1}{8}\frac{1}{16}, \; 1\frac{1}{4}\frac{1}{16}, \; 1\frac{1}{8}\frac{1}{16}, \; 1\frac{1}{16}, \; \frac{1}{2}\frac{1}{4}\frac{1}{8}\frac{1}{16}, \; \frac{1}{2}\frac{1}{4}\frac{1}{16}, \; \frac{1}{2}\frac{1}{8}\frac{1}{16}, \; \frac{1}{2}\frac{1}{16}, \; \frac{1}{4}\frac{1}{8}\frac{1}{16} \; \text{hekats}$$

This explains the solution for solving the arithmetic progression.

Problem KPQV,3

There is a lot of discussion in the literature from many mathematicians and historians pertaining to this problem.

The translation of this problem, according to R. J. Gillings (1972) follows.

<table>
<tr><td>**Column 11**</td><td>**Column 12**</td></tr>
<tr><td>$\backslash 1 \rightarrow \frac{1}{3}\frac{1}{12}$</td><td>110</td></tr>
<tr><td>$2 \rightarrow \frac{2}{3}\frac{1}{6}$</td><td>$\backslash 13\frac{2}{3}\frac{1}{12}$</td></tr>
<tr><td>$4 \rightarrow 1\frac{2}{3}$</td><td>$\backslash 12\frac{2}{3}\frac{1}{6}\frac{1}{12}$</td></tr>
<tr><td>$\backslash 8 \rightarrow \frac{2}{3}\frac{1}{12}$</td><td>$\backslash 12\frac{1}{12}$</td></tr>
<tr><td></td><td>$\backslash 11\frac{1}{6}\frac{1}{12}$</td></tr>
<tr><td></td><td>$\backslash 10\frac{1}{3}\frac{1}{12}$</td></tr>
<tr><td></td><td>$\backslash 9\frac{1}{3}\frac{1}{6}\frac{1}{12}$</td></tr>
<tr><td></td><td>$\backslash 8\frac{2}{3}\frac{1}{12}$</td></tr>
<tr><td></td><td>$\backslash 7\frac{2}{3}\frac{1}{6}\frac{1}{12}$</td></tr>
<tr><td></td><td>$\backslash 7\frac{1}{12}$</td></tr>
<tr><td></td><td>$\backslash 6\frac{1}{6}\frac{1}{12}$</td></tr>
</table>

The sum of these 10 terms is 100. However, if we continue subtracting $\frac{1}{2}\frac{1}{3}$, the next 2 terms will be $5\frac{1}{3}\frac{1}{12}$ and $4\frac{1}{3}\frac{1}{6}\frac{1}{12}$. The sum of the twelve terms, including the last two terms will be 110. Therefore, the number of terms in the arithmetic progression is twelve. We can state the problem as follows:

The sum of 12 terms of an arithmetic progression is 110 and the common difference is $\frac{1}{2}\frac{1}{3}$. Find the terms.

To analyze the problem:

$$n = 12 \quad \text{and } d = \frac{1}{2}\frac{1}{3}, \quad \text{then } \frac{d}{2} = \frac{1}{3}\frac{1}{12}, \quad S = 110$$

The scribe found the last term with the same procedure which was used in Problem 64 (RMP), as was done in the previous problem.

$$\mathcal{L} = \frac{S}{n} + (n-1)\frac{d}{2} = \frac{110}{12} + \left(\frac{1}{3}\frac{1}{12}\right) \times 11 = 9\frac{1}{6} + \frac{55}{12}$$
$$= 9\frac{1}{6} + 4\frac{7}{12} = 13\frac{9}{12} = 13\frac{8}{12}\frac{1}{12} = 13\frac{2}{3}\frac{1}{12}$$

Looking to Column 11, it is a table of multiplication of $i\left(\frac{d}{2}\right)$ from 1, 2, 4, and 8 and a mark on 1 and 8. This table helped the scribe to find the value of $id = 2i\left(\frac{d}{2}\right)$ for $i = 1$ to 12. However, the scribe was interested in finding $i = 1 + 8 = 9$ and calculated from $\mathcal{L}_{12}$ to $\mathcal{L}_3$, which is $\mathcal{L} - 9id$, and the sum was 100, but if we continue to calculate $\mathcal{L}_2$ and $\mathcal{L}_1$, the total sum becomes 110.

$$\mathcal{L}_n = L$$
$$\mathcal{L}_{n-i} = L - id \qquad i = 1, 2, 3, \dots , 12$$

We can use the table, Column 11, to get the values of $\text{id} = 2i\frac{d}{2}$

$$\mathcal{L}_{12} = 13\frac{2}{3}\frac{1}{12}$$

$$\mathcal{L}_{11} = 13\frac{2}{3}\frac{1}{12} - \frac{2}{3}\frac{1}{6} = 12\frac{2}{3}\frac{1}{6}\frac{1}{12}$$

$$\mathcal{L}_{10} = 13\frac{2}{3}\frac{1}{12} - 2\left(\frac{2}{3}\frac{1}{6}\right) = 12\frac{1}{12}$$

$$\mathcal{L}_9 = 13\frac{2}{3}\frac{1}{12} - 3\left(\frac{2}{3}\frac{1}{6}\right) = 11\frac{1}{6}\frac{1}{12}$$

$$\mathcal{L}_8 = 13\frac{2}{3}\frac{1}{12} - 4\left(\frac{2}{3}\frac{1}{6}\right) = 10\frac{1}{3}\frac{1}{12}$$

$$\mathcal{L}_7 = 13\frac{2}{3}\frac{1}{12} - 5\left(\frac{2}{3}\frac{1}{6}\right) = 9\frac{1}{3}\frac{1}{6}\frac{1}{12}$$

$$\mathcal{L}_6 = 13\frac{2}{3}\frac{1}{12} - 6\left(\frac{2}{3}\frac{1}{6}\right) = 8\frac{2}{3}\frac{1}{12}$$

$$\mathcal{L}_5 = 13\frac{2}{3}\frac{1}{12} - 7\left(\frac{2}{3}\frac{1}{6}\right) = 7\frac{2}{3}\frac{1}{6}\frac{1}{12}$$

$$\mathcal{L}_4 = 13\frac{2}{3}\frac{1}{12} - 8\left(\frac{2}{3}\frac{1}{6}\right) = 7\frac{1}{12}$$

$$\mathcal{L}_3 = 13\frac{2}{3}\frac{1}{12} - 9\left(\frac{2}{3}\frac{1}{6}\right) = 6\frac{1}{6}\frac{1}{12}$$

$$\mathcal{L}_2 = 13\frac{2}{3}\frac{1}{12} - 10\left(\frac{2}{3}\frac{1}{6}\right) = 5\frac{1}{3}\frac{1}{12}$$

$$\mathcal{L}_1 = 13\frac{2}{3}\frac{1}{12} - 11\left(\frac{2}{3}\frac{1}{6}\right) = 4\frac{1}{3}\frac{1}{6}\frac{1}{12}$$

$$\text{TOTAL:} \quad 110$$

The scribe may use the table to calculate 'id'. For example:

$$\mathcal{L}_8 = \mathcal{L} - 4d = \mathcal{L} - 4\left(2\frac{d}{2}\right) = \mathcal{L} - 2\left(4\frac{d}{2}\right).$$

From the table in Column 11, $\left(4\frac{d}{2}\right) = 1\frac{2}{3}$

$$\therefore \mathcal{L}_8 = \mathcal{L} - 4d = 13\frac{2}{3}\frac{1}{12} - 2\left(1\frac{2}{3}\right) = 13\frac{2}{3}\frac{1}{12} - 3\frac{1}{3} = 10\frac{1}{3}\frac{1}{12}$$

$$\mathcal{L}_7 = \mathcal{L} - 5d = 13\frac{2}{3}\frac{1}{12} - 5\left(2\frac{d}{2}\right).$$

Using the table in Column 11:

$$\mathcal{L}_7 = 13\frac{2}{3}\frac{1}{12} - 4\frac{1}{6} = 9\frac{1}{3}\frac{1}{6}\frac{1}{12}$$

$\therefore$ Column 11 was a table to help calculate 'd' and the reason the scribe wrote only the ten terms was because he may have been interested in the terms from $\mathcal{L}_{12}$ to $\mathcal{L}_3$.

Note: This is the only problem which shows each detailed calculation.

Problem 40 (RMP)

Divide 100 loaves among five men in such a way that the shares received shall be in an arithmetical progression such that $\frac{1}{7}$ of the sum of the largest three shares shall be equal to the sum of the smallest two shares. What is the difference of the shares?

The scribe used a false position by assuming a $= 1$.

Find 'd' which satisfies the condition, which was $5\frac{1}{2}$.

The way he gets $5\frac{1}{2}$ is as follows:

If the progression is 1, $1 + d$, $1 + 2d$, $1 + 3d$, $1 + 4d$, then he solved the equation as

$$1 + 2d + 1 + 3d + 1 + 4d = 7\,[1 + 1 + d].$$
$$3 + 9d = 14 + 7d; \quad 2d = 11 \ \ d = \longrightarrow 5\frac{1}{2}.$$

Then the progression will be: 23, $17\frac{1}{2}$, 12, $6\frac{1}{2}$, 1 (Which totals 60.)

$$60 \times 1\frac{2}{3} = 100.$$

Multiply the previous progression by $1\frac{2}{3}$:

23 becomes $38\frac{1}{3}$ loaves; $17\frac{1}{2}$ becomes $29\frac{1}{6}$ loaves; 12 becomes 20 loaves;

$6\frac{1}{2}$ becomes $10\frac{2}{3}\frac{1}{6}$ loaves; and 1 becomes $1\frac{2}{3}$ loaves.

Total: 100 loaves.

Problem 1 (P. British 10520; Demotic Mathematical Papyri)

This contains an arithmetic progression a_i and is stated as follows:

Determine the sum of the arithmetic progression, 1 to 10, and the sum of this same progression, 1 to 10, in the second order (which is the sum of the triangular numbers from 1-10)

$$S_1 = 1 + 2 + 3 + 4 + 5 + 6 + 7 + 8 + 9 + 10 = \sum_{i=1}^{10} i$$

$$S_2 = 1 + 3 + 6 + 10 + 15 + 21 + 28 + 36 + 45 + 55 = \sum_{a_1}^{a_n} \sum_{i=1}^{n} i$$

To solve the first problem, the scribe used the formula:

$$S_1 = \frac{n(n+1)}{2} = \frac{n^2+n}{2} \qquad a = 1 \qquad n = 10$$

$$10 \times 10 = 100$$

$$n^2 + n = 100 + 10 = 110$$

$$S_1 = \frac{n^2+n}{2} = \frac{110}{2} = 55$$

For the other progression, the scribe applied the formula:

$$S_2 = \left(\frac{n+2}{3}\right)\left(\frac{n^2+n}{2}\right) = \left(\frac{n+2}{3}\right) \sum_{i=1}^{10} i$$

The proof is as follows:

The 'i' term for the progression $= a_i = \sum_{i=1}^{i} i = \frac{i(i+1)}{2} = \frac{(i^2+i)}{2} = \frac{1}{2}(i^2 + i)$

$$S_2 = \sum_{a=1}^{a_n} a_n = \frac{1}{2}\left[\sum_{i=1}^{n} i^2 + \sum_{i=1}^{n} i\right] = \frac{1}{2}\left[\frac{n(n+1)(2n+1)}{6} + \frac{n(n+1)}{2}\right]$$

$$= \frac{1}{2}\left[\frac{n(n+1)(2n+1+3)}{6}\right] = \frac{1}{2}\frac{n(n+1)(2n+4)}{6}$$

$$= \frac{1}{2}\left[\frac{2n(n+1)(2n+2)}{6}\right] = \frac{n(n+1)(n+2)}{6}$$

The author's solution $= \frac{(n+2)}{3}\frac{n(n+1)}{2} = \left(\frac{n+2}{3}\right)\sum_{i=1}^{10} i = \left(\frac{10+2}{3}\right)(55) = 220.$

The geometric progression is a geometric sequence with a sequence of numbers in which each succeeding term is obtained by multiplying the preceding term by the same number. That number is called the ratio or common ratio of the geometric progression, and if the first term is 'a' and the ratio is 'r', then the sequence takes the form $a, ar, ar^2, \ldots, ar^{n-1}$

and 'n' term $L = ar^{n-1}$

and if $r \neq 1$

the sum of the first 'n' term $s_n = \frac{a(r^n-1)}{r-1}$

The ancient Egyptians discovered that any natural number can be expressed as the sum of some terms of the geometric sequence 1, 2 ,4, 8, ... , and the methods of performing multiplication by continuous doubling would make them interested in the above sequence and other geometric sequences.

The Egyptians developed their own formula for the special case when a = r.

The sequence takes the form $r, r^2, r^3, r^4, \ldots r^n$

$L = r^n$ and a = r

$$s_n = r\left(\frac{(L-1)}{r-1}\right) = r(s_{n-1} + 1)$$

Problem 79 (RMP)

There are seven houses and in each house, there are seven cats. Each cat kills seven mice, and each mouse has eaten seven ears of spelt. Each spelt will produce seven hekats of grain. How much grain has been saved by the cats?

The proof: find the sum of progression $7, 7^2, 7^3, 7^4, 7^5$

The scribe places before the successive power of 7, the word houses, cats, mice, spelt, and hekat.

The scribe solved the problem by finding $L = r^5 = 16,807$.

Since r = 7, apply the formula $s_n = r\left(\frac{(L-1)}{r-1}\right) = \frac{r\,(r^{n-1})}{r-1} = \frac{r(L-1)}{r-1}$ $\quad L = r^n$.

He found that $\frac{(L-1)}{r-1} = \frac{16807-1}{6} = \frac{16806}{6} = 2801$.

Then multiply $r\left(\frac{(L-1)}{r-1}\right) = 7 \times 2801 = 19,607$ hekats.

The first column is the multiplication of 7 by 2801.

$$7 \begin{cases} 2801 \\ 5602 \\ \underline{11,204} \end{cases}$$

The total is 19,607 hekats.

The proof was in the second column (proving the formula he used).

Houses	7
Cats	49
Mice	343
Spelt	2401
Hekat	16,807
Total	19,607

The scribe used the formulas

$s_n = r(s_{n-1} + 1)$ to show that

$$s_{n-1} = \frac{r(r^{n-1})-1}{r-1} = \frac{r^n - r}{r-1}$$

$$(s_{n-1} + 1) = \frac{r^n - r + r - 1}{r-1} = \frac{(r^n-1)}{r-1} = \frac{L-1}{r-1}; \; L = r^n$$

$$r(s_{n-1} + 1) = \frac{r(r^n-1)}{r-1} = \frac{r(L-1)}{r-1} = s_n$$

ALGEBRAIC RULES USED BY ANCIENT EGYPTIANS IN SOLVING PROBLEMS

If $\dfrac{x}{p_1} = \dfrac{y}{p_2}$ then:

1) $\dfrac{y}{x} = \dfrac{p_2}{p_1}$

2) $\dfrac{y-x}{x} = \dfrac{p_2-p_1}{p_1}$

3) $y = \dfrac{(p_2-p_1)}{p_1} x + x$

4) $y = \left(\dfrac{p_2}{p_1}\right) x$

5) $x = \left(\dfrac{p_1}{p_2}\right) y$

6) If $\dfrac{ax}{at} = \dfrac{m}{n}$ then $\dfrac{x}{t} = \dfrac{m}{n}$ and $x = \left(\dfrac{m}{n}\right) t$

7) If $ax + b = c$ then $x = \dfrac{1}{a}(c - b)$ and $b = 0$ then $x = \dfrac{c}{a}$

8) If $x^2 + y^2 = a^2$ and $x = ky$ then $y = \dfrac{a}{\sqrt{k^2+1}}$ and $x = k\left(\dfrac{a}{\sqrt{k^2+1}}\right)$

9) If $xy = a$ and $x = ky$ then $x = \sqrt{ak}$ and $y = \sqrt{\dfrac{a}{k}}$

10) If $x^2 = b^2 + (x - a)^2$ then $x = \dfrac{b^2+a^2}{2a}$

11) If $x^2 + y^2 = d^2$ and $xy = a$ then $x = \dfrac{1}{2}\left(\sqrt{d^2 + 2a} + \sqrt{d^2 - 2a}\right)$

$$\text{and} \quad y = \dfrac{1}{2}\left(\sqrt{d^2 + 2a} - \sqrt{d^2 - 2a}\right)$$

12) If $hw = a$ $\qquad$ (1)

$\quad$ $(h - x)w = a_1$ $\qquad$ (2)

$\quad$ $(h - x)(w + y) = a$ $\qquad$ (3)

$\quad$ then $y = \dfrac{a - a_1}{h - x}$ $\qquad$ or $\qquad$ $y = \dfrac{xw}{h - x}$

13) If $hw = a$ $\qquad$ (1)

$\quad$ and $(h + x)(w - y) = a$ $\qquad$ (2)

$\quad$ $h(w - y) = a_1$ $\qquad$ (3)

$\quad$ then $x = \dfrac{a - a_1}{w - y} = \dfrac{hy}{w - y}$

14) If a, $a + d$, $a + 2d$, ..., $a + (n-1)d$ is an arithmetic progression, 'a' is the first term and 'd' is the common difference; then the average term $= \dfrac{S_n}{n}$; then l (larger term) $= \dfrac{S_n}{n} + (n - 1)\dfrac{d}{2}$.

$S_n = $ the sum of the arithmetic progression.

$$S_n = \frac{n}{2}(a + L) \qquad = \frac{1}{2}n[2a + (n - 1)d]$$

15) $\quad S_n = \displaystyle\sum_{i=1}^{n} i = \dfrac{n^2 + n}{2} \qquad i = 1, 2, 3, \ldots n$

16) $\quad S_n = \displaystyle\sum_{a_1}^{a_n} \sum_{i=1}^{n} i = \Sigma \dfrac{(i^2 + i)}{2} = \left(\dfrac{n+2}{3}\right)\left(\dfrac{n^2 + n}{2}\right) = \dfrac{n + 2}{3}\displaystyle\sum_{i=1}^{n} i$

17) If $r, r^2, r^3, \ldots r^n$ then $s_n = r\left(\dfrac{L-1}{r-1}\right)$ and $L = r^n$

$$s_n = r(s_{n-1} + 1)$$

Chapter 9

Egyptian Measures

The ancient Egyptian measurements came to us from the Palermo Stone and various mathematic papyri. The Ahmes Papyrus (RMP), **Problems 43-55,** and Mosco Papyrus, **Problems 6, 7, and 10** have to do with rectangles, triangles, trapezoids, and circles. The base, side, and diameter are expressed by cubits and khets; areas are expressed by square cubits, cubit strips, and setats. We learn the relationships from the Reisner Papyrus. In the Demotic Papyri, Cairo Papyrus (**Problems 36 and 37**), and the P. British Museum, the royal cubit was used, and it is called the divine cubit or Cubit of Thoth. Length was measured by cubit rods. Some of these cubit rods were found in the tombs of officials such as Maja, Kha, and the treasure of King Tutankhamen. The cubit is 52.5 cm long and is divided into 28 fingers from the left to the right. The fingers are further subdivided into fractions of fingers from right to left. The first finger divides into two parts, the second into three parts, and so on. The measurement could be made to finger fractions with any denominator of 2 through 16; the division $\frac{1}{16}$ finger was equal to $\frac{1}{448}$ of a royal cubit, shown by the following figure. This cubit rod is from the Turin Museum and shows some of the measurements that were used.

For longer distances, such as land measurements, the ancient Egyptians used rope with knots tied at regular intervals. Even though rods and ropes were used, they used the body part to name the linear unit of measurement. The most used linear units were the cubit and the khet.

The following was taken from the problems of different Egyptian papyri. The author added the English measures. He placed emphasis on those measurements used in the ancient papyri for solving the problems.

The following table shows some of the linear units used.

LINEAR UNITS

Name	Egyptian Name	Equivalent Egyptian Values	Metric/U.S. Equivalents
Finger	*djeba*	1 finger = 1/4 palm	18.75 mm .738 in.
Palm	*shesep*	1 palm = 4 fingers	7.5 cm 2.95 in.
Hand	*Drt*	1 hand = 5 fingers	9.38 cm 3.69 in.
Short cubit	*meh nedjes*	1 short cubit = 6 palms = 24 fingers	45 cm 17.72 in.
Royal cubit	*meh niswt*	1 royal cubit = 7 palms = 28 fingers	52.5 cm 20.67 in.

Name	Egyptian Name	Equivalent Egyptian Values	Metric/U.S. Equivalents
Pole	*Nbiw*	1 nbiw = 6 hands = 8 palms = 32 fingers =2 ft.	60 cm 23.62 in.
Rod of cordnwh [or khet]	*nwh [or khet]*	1 rod of cord = 100 royal cubits	52.5 m 172.24 ft.
River measure	*Iteru*	1 iteru = 20,000 cubits	10.5 km 6.52 miles

The River Measure unit is presented to reflect one day of travel towing a boat up the river.

The table shows the relationship between the metric and U.S. equivalents.

1.61 km =   1 mile

5,280 feet =   1 mile

1,760 yards = 1 mile

Additionally: 1 meter =   1 m = 39.37 inches

1 centimeter = 1 cm = .3937 inches

1 millimeter = 1 mm =.03937 inches

In the Demotic Papyri, the cubit = royal cubit = 7 palms. In **Problems 36 and 37** in the CMP, and **Problems 2 and 3** in the PM 10399, they call it the royal cubit or divine cubit, and in **Problem 2** it also is called the Cubit of Thoth.

AREA UNITS OF MEASUREMENT

The area measurements came to us from the gift of land recorded in the Palermo Stone and other mathematic papyri expressed in terms of square cubits and square khets, which was called setat, which equaled 10,000 square cubits. The setat could be divided into strips called cubit strips which is khet or 100 cubits long and one cubit wide. The following figure shows some of the area units.

In the last problem in PBM 10520, the fraction of the Aroura was mentioned, which is strictly a measure of the area that was used linearly to give the dimensions of the field. In other problems, there are also land cubits where a land cubit equals a cloth cubit equals 100 square cubits, or $\frac{1}{100}$ of an Aroura. The Aroura equals the setat = 10,000 square cubits.

UNITS OF AREA

Name	Egyptian Name	Equivalent Egyptian Values	Metric/U.S. Equivalents
S3	*s3*	1/8 st3t (Aroura of 1 sq. khet)	1250 sq cubits 344.53125 sq meters 3512.84 sq. feet = .08 acre
hsb	*Hsb*	1/4 st3t (Aroura of 1 sq. khet)	2500 sq cubits 689.0625 sq meters 7,025.68 sq. feet = .16 acre

Name	Egyptian Name	Equivalent Egyptian Values	Metric/U.S. Equivalents
rmn	*rmn*	1/2 st3t (Aroura of 1 sq. khet)	5000 sq cubits 1378.125 sq meters 14,051.36 sq. feet = .32 acre
Setat (setjat)	*setat*	1 square khet = 10,000 square cubits	2,756½ sq. meters 2,965.94 sq. feet
h3-t3	*Kha*	1000 of land	10 arouras, 100,000 sq cubits, 27562.5 sq m. 296,572.5 sq. feet
Ta	*ta* *cubit strip*	100 square cubits = 1/100 setat	27.565 square meters 296.594 sq. feet
Shoulder (Remen)	*remen*	1/2 ta = 50 square cubits	13.7 square meters 147.41 sq. feet
Heseb	*heseb*	1/2 remen = 25 square cubits	6.8 square meters 73.16 sq. feet

This figure illustrates some of the units of area used by the ancient Egyptians.

Aroura is also a Greek measure which is equivalent to one square khet = 10,000 square cubits = 1 setat.

UNITS OF VOLUME AND CAPACITY

In the RMP mathematical papyrus, some of the problems dealt with volume and capacity. For example, **Problems 41-46** dealt with the volumes of rectangular and cylindrical granaries, and in the Cairo Papyrus, **Problems 39 and 40** dealt with the volume of the pyramids. The unit used in these problems and other volume problems in other papyri were cubic cubits, khar, hekat, and quadruple hekat (heqat). The hekat was divided into 320 parts called ro, but the Egyptian also used a fraction of a hekat which is a power of $\frac{1}{2}$, down to $\frac{1}{64}$, and $\frac{1}{64}$ of a hekat is 5 ro. The following table illustrates some of the volume and capacity units used by the ancient Egyptians.

Note: 20 khar = 100 quadruple hekats.

UNITS OF VOLUME AND CAPACITY

Name	Egyptian Name	Equivalent Egyptian Values	Metric/U.S. Equivalents
Deny	*deny*	1 cubic cubit	144.75 liters
Khar (sack)	*khar*	$\frac{2}{3}$ cubic cubits 20 hekat (Middle Kingdom) 16 hekat (New Kingdom)	96.5 liters (Middle Kingdom) 23.35 gr. 76.8 liters (New Kingdom) 20.284 gr. 2/3 cubed cubit 6,266.92 cubic inches
quadruple heqat (hekat)	*hekat-fedw*	4 hekat = 40 hinu	19.2 liters = 1,168 96 cubic inches = 5.071 gallons

Name	Egyptian Name	Equivalent Egyptian Values	Metric/U.S. Equivalents
Double Heqat (hekat)	*hekaty*	2 hekat = 20 hinu	9.6 liters = 584.48 cubic inches = 2.535 gallons
Heqat (hekat) (barrel)	*hekat*	$\frac{1}{30}$ cubic cubits = 10 hinu	4.8 liters = 292.24 cubic inches = 1.267 gallons
Hinu (jar) (hin) (hini)	*hnw*	$\frac{1}{300}$ cubic cubits 1/10 hekat = 32 ro	.48 liters = 292.24 cubic inches = 1.267 gallons
Dja	*dja*	5/8 hinu = 20 ro	.3 liters
Ro	*r*	1/320 hekat	.015 liters = .00396 gallons

Additionally, 1 liter = .26412 gallons.

ANCIENT EGYPTIAN MEASURES COMPARED TO MODERN MEASURES

Ro = 15 ml. This is very close to a modern tablespoon measure of 14.79 ml.

In the medicine of ancient Egypt, the most common unit is the hini, which is also the name of the vessel that had a standard volume equal to a measuring cup.

In text, the amount of certain ingredients is saved in hini and ro. Other measures used for grain were $\frac{1}{2}$, $\frac{1}{4}$, $\frac{1}{8}$, $\frac{1}{16}$, $\frac{1}{32}$, and $\frac{1}{64}$ hekat. (Strouhal, Vachala, &Vymazalova, 2014, p.103.)

UNITS OF WEIGHT

Weight was measured in terms of deben, kites, and shematy (shaty). **Problem 62** in the Rhind Mathematical Papyrus expressed the relationship between the deben and the shematy when given one deben of gold as equal to 12 shematy of gold. The deben changed from 13.6 grams in the Old and Middle Kingdoms to 91 grams during the New Kingdom. The following are some of the weights used.

UNITS OF WEIGHT

Name	Egyptian Name	Equivalent Egyptian Values	Metric/U.S. Equivalents
Deben	*dbn*		13.6 grams in the Old Kingdom and Middle Kingdom. = .476 ounces 91 grams during the New Kingdom = 3.185 ounces
Kite	*qd.t*	1/10 of a deben	1.36 grams in the Old Kingdom and Middle Kingdom = .0476 oz. 9.1 grams during the New Kingdom = .3185 ounces
Shematy (shaty)	*shȝts*	1/12 of a deben	1.333 grams in the Old Kingdom and Middle Kingdom = .0465 oz. 7.58 grams during the New Kingdom = .265 ounces

Not all measurements were units; Ro, for example, were unit fractions which the Egyptian used instead of decimals or other fractions. There are tables of unit fractions in some of the papyri, such as the Rhind Mathematical Papyri and section K in the Reisner Papyrus, to help divide units into parts. The following is an example of a **TABLE OF DIVISION 1-9 BY 10**, as listed in the RMP.

Number	1ˢᵗ Quotient	2ⁿᵈ Quotient	3ʳᵈ Quotient
1	$\frac{1}{10}$		
2	$\frac{1}{5}$		
3	$\frac{1}{5}$	$\frac{1}{10}$	
4	$\frac{1}{3}$	$\frac{1}{15}$	
5	$\frac{1}{2}$		
6	$\frac{1}{2}$	$\frac{1}{10}$	
7	$\frac{2}{3}$	$\frac{1}{30}$	
8	$\frac{2}{3}$	$\frac{1}{10}$	$\frac{1}{30}$
9	$\frac{2}{3}$	$\frac{1}{5}$	$\frac{1}{30}$

This table was listed before the first six problems in the RMP to help the scribe solve the problems. Similar tables were listed in section K of the Reisner Papyrus to help the builder construct the different chambers of the temple. This table can be used to take $\frac{1}{10}$ of any number which is not a multiple of 10.

RELATIONSHIP BETWEEN FRACTIONS OF HEKATS AND HINU

There is also a table to express the hekat and its parts in terms of hinu and ro, such as is found in the following problems.

Problem 80 (RMP)

Express the hekat in terms of hinu. The scribe's answer:

1 hekat makes 10 hinu

$\frac{1}{2}$ hekat makes 5 hinu

$\frac{1}{4}$ hekat makes $2\frac{1}{2}$ hinu

$\frac{1}{8}$ hekat makes $1\frac{1}{4}$ hinu

$\frac{1}{16}$ hekat makes $\frac{1}{2}\frac{1}{8}$ hinu

$\frac{1}{32}$ hekat makes $\frac{1}{4}\frac{1}{16}$ hinu

$\frac{1}{64}$ hekat makes $\frac{1}{8}\frac{1}{32}$ hinu

The scribe found the values of $1, \frac{1}{2}, \frac{1}{4}, \frac{1}{8}, \frac{1}{16}, \frac{1}{32}$, and $\frac{1}{64}$ of the hekat in terms of hinu.

Problem 81 (RMP)

There is a full table of fractional parts of a hekat in terms of the hinu, ro, and a part of the hekat. It follows.

$\frac{1}{2}\,\frac{1}{4}\,\frac{1}{8}$	hekat		makes $8\frac{1}{2}\frac{1}{4}$ hinu	
$\frac{1}{2}\,\frac{1}{4}$	hekat		makes $7\frac{1}{2}$ hinu	
$\frac{1}{2}\,\frac{1}{8}\,\frac{1}{32}$	hekat	$3\frac{1}{3}$ ro	makes $6\frac{2}{3}$ hinu;	it is $\frac{2}{3}$ hekat
$\frac{1}{2}\,\frac{1}{8}$	hekat		makes $6\frac{1}{4}$ hinu;	it is $\frac{1}{2}\frac{1}{8}$ hekat
$\frac{1}{4}\,\frac{1}{8}$	hekat		makes $3\frac{1}{2}\frac{1}{4}$ hinu;	it is $\frac{1}{4}\frac{1}{8}$ hekat
$\frac{1}{4}\,\frac{1}{16}\,\frac{1}{64}$	hekat	$1\frac{2}{3}$ ro	makes $3\frac{1}{3}$ hinu;	it is $\frac{1}{3}$ hekat
$\frac{1}{8}\,\frac{1}{16}$	hekat	4 ro	makes 2 hinu;	it is $\frac{1}{5}$ hekat
$\frac{1}{8}\,\frac{1}{32}$	hekat	$3\frac{1}{3}$ ro	makes $1\frac{2}{3}$ hinu;	it is $\frac{1}{6}$ hekat
$\frac{1}{8}\,\frac{1}{16}$	hekat	4 ro	makes 2 hinu;	it is $\frac{1}{5}$ hekat
$\frac{1}{16}\,\frac{1}{32}$	hekat	2 ro	makes 1 hinu;	it is $\frac{1}{10}$ hekat
$\frac{1}{32}\,\frac{1}{64}$	hekat	1 ro	makes $\frac{1}{2}$ hinu;	it is $\frac{1}{20}$ hekat
$\frac{1}{64}$	hekat	3 ro	makes $\frac{1}{4}$ hinu;	it is $\frac{1}{40}$ hekat
$\frac{1}{16}$	hekat	$1\frac{1}{3}$ ro	makes $\frac{2}{3}$ hinu;	it is $\frac{1}{15}$ hekat
$\frac{1}{32}$	hekat	$\frac{2}{3}$ ro	makes $\frac{1}{3}$ hinu;	it is $\frac{1}{30}$ hekat
$\frac{1}{64}$	hekat	$\frac{1}{3}$ ro	makes $\frac{1}{6}$ hinu;	it is $\frac{1}{60}$ hekat
$\frac{1}{2}\,\frac{1}{4}$	hekat		makes $7\frac{1}{2}$ hinu;	it is $\frac{1}{2}\frac{1}{4}$ hekat
$\frac{1}{2}\,\frac{1}{4}\,\frac{1}{8}$	hekat		makes $8\frac{1}{2}\frac{1}{4}$ hinu;	it is $\frac{1}{2}\frac{1}{4}\frac{1}{8}$ hekat
$\frac{1}{2}\,\frac{1}{8}\,\frac{1}{32}$	hekat	$3\frac{1}{3}$ ro	makes $6\frac{2}{3}$ hinu;	it is $\frac{2}{3}$ hekat
$\frac{1}{4}\,\frac{1}{16}\,\frac{1}{64}$	hekat	$1\frac{2}{3}$ ro	makes $3\frac{1}{3}$ hinu;	it is $\frac{1}{3}$ hekat

This table in **Problem 80** is not the full table. The rest is explained in **Problem 84**. (Chace, 1927, p. 113.) Some of the errors made by the scribe were corrected by Chace and this author.

In this table, the scribe found the value of

$\frac{1}{3}, \frac{2}{3}, \frac{1}{6}, \frac{1}{5}, \frac{1}{10}, \frac{1}{15}, \frac{1}{20}, \frac{1}{30}, \frac{1}{40},$ and $\frac{1}{60}$ of the hekat in terms of hinus and ros.

HEKAT PROBLEMS

Problem 35 (RMP)

I have got 3 times in hekat measures, my $\frac{1}{3}$ is added to me; I have returned having filled the hekat measure.

The Solution

Assume 1

$$1 \longrightarrow 1$$
$$2 \longrightarrow 2$$
$$\frac{1}{3} \longrightarrow \frac{1}{3}$$

Get 1 by multiplying by $3\frac{1}{3}$

$$1 \longrightarrow 3\frac{1}{3}$$

$$\frac{1}{5}\frac{1}{10} \left\langle \begin{array}{ccc} \frac{1}{10} & \longrightarrow & \frac{1}{3} \\ \frac{1}{5} & \longrightarrow & \frac{2}{3} \end{array} \right\rangle 1$$

The answer is $\frac{1}{5}\frac{1}{10}$ hekat by Egyptian fractions *

The Proof

$$1 \longrightarrow \frac{1}{5}\frac{1}{10}$$
$$2 \longrightarrow \frac{1}{2}\frac{1}{10}$$
$$\frac{1}{3} \longrightarrow \frac{1}{10}$$

The total is 1 hekat.

*__Note__: The scribe found the answer $= \frac{1}{5}\frac{1}{10}$ hekat in Egyptian fractions and to find the answer in ro $= 96$ ro.

Consider the hekat = 320 units (ros).

$$1 \longrightarrow 320 \text{ ro}$$

$$\frac{1}{10} \longrightarrow 32 \text{ ro}$$

$$\frac{1}{5} \longrightarrow 64 \text{ ro}$$

The total = 96 ro.

<u>The Proof</u>

$$1 \longrightarrow 96 \text{ ro}$$

$$2 \longrightarrow 192 \text{ ro}$$

$$\frac{1}{3} \longrightarrow 32 \text{ ro}$$

$$320 = 1 \text{ hekat.}$$

Expressed using the Horus Eye, it makes the grain $\frac{1}{4}\frac{1}{32}\frac{1}{64}$ of a hekat.

$$1 \longrightarrow \frac{1}{4}+\frac{1}{32}+\frac{1}{64} \text{ hekat} + 1 \text{ ro}$$

$$2 \longrightarrow \frac{1}{2}+\frac{1}{16}+\frac{1}{32} \text{ hekat} + 2 \text{ ro}$$

$$\frac{1}{3} \longrightarrow \frac{1}{16}+\frac{1}{32} \text{ hekat} + 2 \text{ ro}$$

$$\frac{1}{2}+\frac{1}{4}+\left(\frac{1}{16}+\frac{1}{16}\right)+\left(\frac{1}{32}+\frac{1}{32}\right)+\frac{1}{32}+\frac{1}{64} \text{ hekat} + 5 \text{ ro}$$

The total $= \frac{1}{2}+\frac{1}{4}+\frac{1}{8}+\frac{1}{16}+\frac{1}{32}+\frac{1}{64}$ hekat + 5 ro = 1 hekat.

From that, the Horus Eye System was written to measure the grain and using ro so as to equal $\frac{1}{320}$ of the hekat, which makes the answer exact.

<u>The Proof</u>

Multiply 96 by $3\frac{1}{2}$; the answer is 320 ro = 1 hekat

Also multiply $3\frac{1}{2}$ by $\frac{1}{2}+\frac{1}{16}+\frac{1}{32}$ hekat + 1 ro

The answer $= \frac{1}{2}+\frac{1}{4}+\frac{1}{8}+\frac{1}{16}+\frac{1}{32}+\frac{1}{64}$ hekat + 5 ro = 1 hekat, which is in two parts: hekat and ro.

EXAMPLES OF HEKAT PROBLEMS EXPRESSED IN HINU

Problem 80 (RMP)

Since the hekat is 10 hinu, then:

Example 1:

$$\tfrac{1}{2} \text{ hekat} = \tfrac{1}{2} \times 10 = 5 \text{ hinu}$$

Example 2:

$$\tfrac{1}{64} \text{ hekat} = \tfrac{1}{64} \times 10 = \tfrac{10}{64} = \tfrac{8+2}{64} = \tfrac{8}{64} + \tfrac{2}{64} = \tfrac{1}{8}\,\tfrac{1}{32} \text{ hinu.}$$

In general, 1 hekat = 10 hinu, and 1 hinu = $\tfrac{1}{10}$ hekat.

From Problem 81 (RMP)

(Expressed hekat in terms of hinu)

$$\tfrac{1}{2}\,\tfrac{1}{4}\,\tfrac{1}{8} \text{ hekat} = \left(\tfrac{1}{2} + \tfrac{1}{4} + \tfrac{1}{8}\right) 10 = 5 + 2\tfrac{1}{2} + 1\tfrac{1}{4}$$

$$= 8\tfrac{1}{2}\,\tfrac{1}{4} \text{ hinu.}$$

DIVISION OF HEKAT

By dividing a hekat by 'n', Ahmes, in 29 cases, used the theory of arithmetic remainders. He scaled the hekat unit by $\tfrac{64}{64}$. Then he applied the theory of arithmetic remainders:

$$\left(\tfrac{64}{64}\right) / n = \tfrac{Q}{64} + \tfrac{R}{n(64)} \qquad \tfrac{1}{64} < n < 64$$

'Q' is the quotient and 'R' is the remainder if dividing 64 by 'n'

The first part, $\dfrac{Q}{64}$ is expressed as a series of binary fractions of hekat.

The scribe multiplied the scaled second part, $\dfrac{R}{n(64)} = \dfrac{R}{n}\left(\dfrac{1}{64}\right)\left(\dfrac{5}{5}\right)$

1 hekat = 320 ro; then

$$\dfrac{5R}{n}\left(\dfrac{1}{320}\right)(320) = \dfrac{5R}{n}\ \text{ro.}$$

The remainder equation:

$$\left(\dfrac{64}{64}\right)/n = \dfrac{Q}{64} + \dfrac{R}{n(64)}$$

Multiply $\dfrac{R}{n(64)}$ by 320   1 hekat = 320 ro

$$\dfrac{R}{n(64)} \times 320 = \dfrac{5R}{n}\ \text{ro}$$

Then the remainder equation is $\left(\dfrac{64}{64}\right)/n = \dfrac{Q}{64} + \dfrac{5R}{n}.$

The scribe writes the first part, $\dfrac{Q}{64}$, in the Horus Eye series (binary fractions), and the second part, $\dfrac{5R}{n}$, in Egyptian fractions.

From Problem 81 (RMP)

Find $\frac{1}{3}$ hekat; divide the hekat by 3.

The Solution

The unit hekat is scaled as $\frac{64}{64}$.

$$\frac{1}{3} \times \frac{64}{64} = \frac{1}{3}\left(\frac{63+1}{64}\right) = \frac{21}{64} + \frac{1 \times 5}{3}$$

$$= \frac{16+4+1}{64} + \left(\frac{5}{3}\text{ro}\right) \qquad (\text{Hekat} = 320 \text{ ro})$$

$$= \left(\frac{1}{4} + \frac{1}{16} + \frac{1}{64}\right) \text{hekat} + \left(1 + \frac{2}{3}\right) \text{ ro.}$$

The answer is a two-part statement; part in hekats and the other in ro.

In general, to divide one hekat by 'n', the scribe followed the modern arithmetic remainder form.

$$\left(\frac{64}{64}\right)/n = \frac{Q}{64} + \frac{R}{n(64)} \text{ hekat} \quad \frac{1}{64} < n < 64$$

'Q' is the quotient and 'R' is the remainder of dividing $\frac{64}{n}$.

From Problem 81 (RMP)

Divide 1 hekat by 5 (express the hekat portion in binary fractions).

The Solution:

$$\left(\frac{64}{64}\right) / 5 = \frac{Q}{64} + \left(\frac{5R}{n}\right) = \frac{Q}{64} \text{ hekat} + \left(\frac{5R}{n}\right) \text{ ro}$$

$$Q = 12 \quad R = 4$$

$$\left(\frac{64}{64}\right) / 5 = \frac{12}{64} \text{ hekat} + \left(\frac{5 \times 4}{5}\right) \text{ ro}$$

$$= \frac{8}{64} + \frac{4}{64} \text{ hekat} + 4 \text{ ro}$$

$$= \frac{1}{8} + \frac{1}{16} \text{ hekat} + 4 \text{ ro.}$$

From Problem 83 (RMP)

If the feed of four geese is one hinu of lower Egyptian grain, the portion of one goose is $\frac{1}{64}$ hekat 3 ro.

The Solution

1 hinu for 4 geese $= \frac{1}{10}$ hekat; for 1 goose $= \frac{1}{40}$ hekat

$\left(\frac{64}{64}\right) / 40 = \frac{Q}{64}$ hekat $+ \left(\frac{5R}{n}\right)$ ro

$Q = 1 \quad R = 24$

$\left(\frac{64}{64}\right) / 40 = \frac{1}{64}$ hekat $+ \frac{5 \times 24}{40}$ ro $= \frac{1}{64}$ hekat $+ 3$ ros of feed per goose.

AKHMIN WOODEN TABLETS (AWT) PROBLEMS

The problems are stated in our Base 10.

The hekat unit is scaled to $\frac{64}{64}$ and was divided by 3, 7, 10, 11, and 13.

The division follows the rule: $\left(\frac{64}{64}\right)/n = \frac{Q}{64}$ hekat $+ \left(\frac{5R}{n}\right) = \frac{Q}{64} + \frac{5R}{n}$.

The scribe used hekat $= 320$ ro and $5R = \frac{1}{64}$ hekat.

The Solution

The scribe of the Akhmin Wooden Tablets, in his solution, used two concepts:

#1: Scaling, which multiplied the numerator and denominator by a number called, in modern times, a practical number. The scribe scaled the fractions which Ahmes used in developing the $\frac{2}{n}$ table. Ahmes also used the arithmetic remainder formula.

$$\text{For example: } \frac{2}{101} = \frac{2\times6}{101\times6} = \frac{12}{606} = \frac{6+3+2+1}{606} = \frac{6}{606} + \frac{3}{606} + \frac{2}{606} + \frac{1}{606}$$

$$= \frac{1}{101} + \frac{1}{202} + \frac{1}{303} + \frac{1}{606}$$

This was $\left(\frac{64}{64}\right)/n = \frac{Q}{64}$ hekat $+ \frac{5R}{n}$ ro.

Q is the quotient and R is the remainder of dividing 64 by 'n'.

The scribe of the AWT was familiar with the hekat.

$$= \frac{1}{2} + \frac{1}{4} + \frac{1}{8} + \frac{1}{16} + \frac{1}{32} + \frac{1}{64} \text{ hekat} + 5 \text{ ro and } 5 \text{ ro} = \frac{1}{64} \text{hekat.}$$

The Proof

To prove that in our Base 10 system:

$$\frac{64}{64 \times 1} = \frac{32+16+8+4+2+1}{64}$$

$$= \frac{32}{64} + \frac{16}{64} + \frac{8}{64} + \frac{4}{64} + \frac{2}{64} + \frac{1}{64} + \frac{1}{64} \text{ hekat.}$$

$$= \frac{1}{2} + \frac{1}{4} + \frac{1}{8} + \frac{1}{16} + \frac{1}{32} + \frac{1}{64} + \frac{1}{64} \text{ hekat.}$$

Note: $\frac{1}{64}$ hekat $= 5$ ro

$$= \frac{1}{2} + \frac{1}{4} + \frac{1}{8} + \frac{1}{16} + \frac{1}{32} + \frac{1}{64} \text{ hekat} + 5 \text{ ro}$$

which is two parts; #1: a binary fraction of the Horus Eye series of hekat.

#2: The second part is a series of Egyptian fractions followed by ro, and $\frac{5R}{n}$ makes 'n' grow by any size by allowing an exact computation each time.

Problem 1

Divide one hekat, by writing as $\frac{1}{2} + \frac{1}{4} + \frac{1}{8} + \frac{1}{16} + \frac{1}{32} + \frac{1}{64} + 5$ ro, by 3 .

The Solution (in our Base 10):

Take 5 ro and divide that by 3. The answer is $\left(1 + \frac{2}{3}\right)$ ro.

The hekat is scaled to $\frac{64}{64}$ and 5 ro $= \frac{1}{64}$; the remainder $= \frac{64}{64} - \frac{1}{64} = \frac{63}{64}$

Then divide $\frac{63}{64}$ by 'n' $= \left(\frac{63}{64}\right) / 3 = \frac{21}{64}$

$21 = 16 + 4 + 1$

16 out of 64 $= \frac{1}{4}$

4 out of 64 $= \frac{1}{16}$

1 out of 64 $= \frac{1}{64}$

The answer $= \frac{1}{4} + \frac{1}{16} + \frac{1}{64}$ hekat $+ \left(1 + \frac{2}{3}\right)$ ro.

The Proof

Multiply $\left(1 + \frac{2}{3}\right)$ ro by 3; the answer is 5 ro.

Multiply $3\left(\frac{1}{4} + \frac{1}{16} + \frac{1}{64}\right)$;

$3 \times \frac{1}{4} = \frac{1}{2} + \frac{1}{4}$

$3 \times \frac{1}{16} = \frac{1}{8} + \frac{1}{16}$

$3 \times \frac{1}{64} = \frac{1}{32} + \frac{1}{64}$

The answer is $\frac{1}{2} + \frac{1}{4} + \frac{1}{8} + \frac{1}{16} + \frac{1}{32} + \frac{1}{64}$ hekat $+ 5$ ro $= 1$ hekat.

Problem 2

Divide $\frac{1}{2} + \frac{1}{4} + \frac{1}{8} + \frac{1}{16} + \frac{1}{32} + \frac{1}{64}$ hekat + 5 ro (or 1 hekat) by 7.

<u>The Solution</u>

Divide 5 ro by 7 = scale $\frac{5}{7} = \frac{10}{14} = \frac{7+2+1}{14}$

$= \left(\frac{7}{14} + \frac{2}{14} + \frac{1}{14} \right)$ ro

$= \left(\frac{1}{2} + \frac{1}{7} + \frac{1}{14} \right)$ ro

Take 5 ro $= \frac{1}{64}$ hekat from $\frac{64}{64}$ hekat $= \frac{63}{64}$ hekat.

$\frac{63}{64 \times 7} = \frac{9}{64} = \frac{8+1}{64} = \frac{8}{64} + \frac{1}{64} = \frac{1}{8} + \frac{1}{64}$

The answer is $\frac{1}{8} + \frac{1}{64}$ hekat $+ \left(\frac{1}{2} + \frac{1}{7} + \frac{1}{14} \right)$ ro.

<u>The Proof</u>

Multiply $7 \left(\frac{1}{2} + \frac{1}{7} + \frac{1}{14} \right) = 3\frac{1}{2} + 1 + \frac{1}{2} = 5$ ro.

Multiply $7 \left(\frac{1}{8} + \frac{1}{64} \right) = \frac{7}{8} + \frac{7}{64} = \frac{4+2+1}{8} + \frac{4+2+1}{64}$

$= \frac{4}{8} + \frac{2}{8} + \frac{1}{8} + \frac{4}{64} + \frac{2}{64} + \frac{1}{64}$

The answer is $\frac{1}{2} + \frac{1}{4} + \frac{1}{8} + \frac{1}{16} + \frac{1}{32} + \frac{1}{64} + 5$ ro $= 1$ hekat.

Problem 3

Divide $\frac{1}{2}+\frac{1}{4}+\frac{1}{8}+\frac{1}{16}+\frac{1}{32}+\frac{1}{64}$ hekat + 5 ro (or 1 hekat) by 10.

The Solution

Divide 5 ro by 10; the answer is $\frac{1}{2}$ ro.

$5 \text{ ro} = \frac{1}{64}$ from $\frac{64}{64} = \frac{63}{64}$

$Q = 6 \quad R = 3$

$\left(\frac{63}{64}\right) / 10 = \frac{Q}{64} + \frac{5R}{n}$

$\left(\frac{63}{64}\right) / 10 = \frac{4+2}{64} \text{ hekat} + \frac{5\times3}{10}\text{ro} = \frac{4}{64} + \frac{2}{64} \text{ hekat} + \frac{15}{10} \text{ ro}$

$$= \frac{1}{16} + \frac{1}{32} \text{ hekat} + \left(1\frac{1}{2} + \frac{1}{2}\right) \text{ro}$$

The answer is $\frac{1}{16} + \frac{1}{32}$ hekat + 2 ro.

The Proof

Multiply 10 by 2 ro = 20 ro = 15 ro + 5 ro

$15 \text{ ro} = \frac{3}{64} \text{ hekat} = \frac{2}{64} + \frac{1}{64} = \frac{1}{32} + \frac{1}{64} \text{ hekat} \quad 5 \text{ ro} = \frac{1}{64} \text{ hekat}$

Multiply 10 by $\frac{1}{16} + \frac{1}{32} = \frac{8+2}{16} + \frac{8+2}{32}$

$$= \frac{8}{16} + \frac{2}{16} + \frac{8}{32} + \frac{2}{32}$$

$$= \frac{1}{2} + \frac{1}{8} + \frac{1}{4} + \frac{1}{16}$$

The answer is $\frac{1}{2} + \frac{1}{4} + \frac{1}{8} + \frac{1}{16} + \frac{1}{32} + \frac{1}{64}$ hekat + 5 ro.

Problem 4

Divide $\frac{1}{2} + \frac{1}{4} + \frac{1}{8} + \frac{1}{16} + \frac{1}{32} + \frac{1}{64} + 5$ ro (or 1 hekat) by 11.

<u>The Solution</u>

Divide 5 by $11 = (\frac{5}{11})$

Take 5 ro $= \frac{1}{64}$ from the hekat $= \frac{64}{64}$; the result is $\frac{63}{64}$.

$$\left(\frac{63}{64}\right) / 11 = \frac{55+8}{64\times11} = \frac{5}{64} \text{ hekat} + \frac{5\times8}{11} \text{ ro}$$

$$= \frac{4+1}{64} + \frac{40}{11} \text{ro}$$

$$= \frac{1}{16} + \frac{1}{64} \text{ hekat} + \frac{40}{11} \text{ ro}$$

The answer is $= \frac{1}{16} + \frac{1}{64}$ hekat $+ \frac{45}{11}$ ro

$$= \frac{1}{16} + \frac{1}{64} \text{ hekat} + 4\frac{1}{11} \text{ ro.}$$

<u>The Proof</u>

Multiply $\frac{5}{11}$ ro by 11; the result is 5 ro

Multiply $\frac{40}{11}$ x11 $= 40$ ro $= \frac{8}{64}$ hekat $= \frac{1}{8}$ hekat   5 ro $= \frac{1}{64}$ hekat

Multiply $11\left(\frac{1}{16} + \frac{1}{64}\right) = \frac{11}{16} + \frac{11}{64}$

$$= \frac{8+2+1}{16} + \frac{8+2+1}{64}$$

$$= \frac{8}{16} + \frac{2}{16} + \frac{1}{16} + \frac{8}{64} + \frac{2}{64} + \frac{1}{64}$$

The answer is $\frac{1}{2} + \frac{1}{4} + \frac{1}{8} + \frac{1}{16} + \frac{1}{32} + \frac{1}{64}$ hekat $+ 5$ ro $= 1$ hekat.

Problem 5

Divide $\frac{1}{2} + \frac{1}{4} + \frac{1}{8} + \frac{1}{16} + \frac{1}{32} + \frac{1}{64}$ hekat + 5 ro (or 1 hekat) by 13.

<u>The Solution</u>

Take 5 ro $= \frac{1}{64}$ and divide by $13 = \frac{5}{13}$ ro.

The hekat unit $= \frac{64}{64}$

The remainder of the hekat $= \frac{64}{64} - \frac{1}{64} = \frac{63}{64}$.

$\left(\frac{63}{64}\right) / 13 = \frac{52+11}{64\times13} = \frac{4}{64}$ hekat $+ \frac{5\times11}{13}$ ro

$\qquad\qquad = \frac{1}{16}$ hekat $+ \frac{55}{13}$ ro

Total ro $= \frac{55}{13} + \frac{5}{13} = \frac{60}{13} = \frac{52}{13} + \frac{8}{13} = 4 + \frac{16}{26}$ ro   Scale $\frac{8}{13} = \frac{16}{26}$

$\qquad\qquad\qquad = 4$ ro $+ \frac{13}{26} + \frac{2}{26} + \frac{1}{26}$ ro

$\qquad\qquad\qquad = 4 + \frac{1}{2} + \frac{1}{13} + \frac{1}{26}$ ro

The answer is $\frac{1}{16}$ hekat $+ \left(4 + \frac{1}{2} + \frac{1}{13} + \frac{1}{26}\right)$ ro.

<u>The Proof</u>

Multiply 13 by $\left(4 + \frac{1}{2} + \frac{1}{13} + \frac{1}{26}\right) = 60$ ro.

Take 5 ro and the result is 55 ro $= \frac{11}{64}$ hekat $= \frac{8+2+1}{64}$

$\qquad\qquad\qquad = \frac{8}{64} + \frac{2}{64} + \frac{1}{64}$

$\qquad\qquad\qquad = \frac{1}{8} + \frac{1}{32} + \frac{1}{64}$

Multiply 13 by $\frac{1}{16} = \frac{13}{16} = \frac{8+4+1}{16} = \frac{8}{16} + \frac{4}{16} + \frac{1}{16} = \frac{1}{2} + \frac{1}{4} + \frac{1}{16}$

The answer is $\frac{1}{2} + \frac{1}{4} + \frac{1}{8} + \frac{1}{16} + \frac{1}{32} + \frac{1}{64}$ hekat + 5 ro.

Looking to the AWT problems and their solutions, it was easy to convert $\dfrac{5R}{n}$ to Egyptian fractions using the method of scaling and the solution by putting the hekat $= \dfrac{1}{2} + \dfrac{1}{4} + \dfrac{1}{8} + \dfrac{1}{16} + \dfrac{1}{32} + \dfrac{1}{64}$ hekat $+ 5$ ro.

Problem 47 (RMP)

Suppose the scribe says, "Let me know what is the result when 100 hekats are divided by 10 and its multiples."

We will choose dividing the 100 hekats by 90.

We scale the 100 hekat by $\dfrac{100 \times 64}{64}$

<u>The Solution</u>

Then $\dfrac{6400}{64 \times 90} = \dfrac{Q}{64} + \dfrac{5R}{n}$

$= \dfrac{71}{64}$ hekat $\dfrac{50}{90}$ ro

$= 1 + \dfrac{7}{64}$ hekat $+ \dfrac{5}{9}$ ro $\quad$ Scale $\dfrac{5}{9} = \dfrac{10}{18} = \dfrac{9+1}{18}$

$= 1 + \dfrac{4+2+1}{64}$ hekat $+ \dfrac{9+1}{18}$ ro

$= 1 + \dfrac{4}{64} + \dfrac{2}{64} + \dfrac{1}{64}$ hekat $+ \dfrac{9+1}{18}$ ro

$= 1 + \dfrac{1}{16} + \dfrac{1}{32} + \dfrac{1}{64}$ hekat $+ \dfrac{1}{2} + \dfrac{1}{18}$ ro.

'n' could be any rational number.

The Proof

Multiply $\left(\frac{1}{2} + \frac{1}{18}\right) \times 90 = 45 + 5 = 50$ ro

Multiply $\left(1 + \frac{1}{16} + \frac{1}{32} + \frac{1}{64}\right) \times 90$

$= 90 + \frac{90}{16} + \frac{90}{32} + \frac{90}{64}$

$= 90 + 10\left(\frac{9}{16} + \frac{9}{32} + \frac{9}{64}\right)$ hekat

$= 90 + 10\left(\frac{8+1}{16} + \frac{8+1}{32} + \frac{8+1}{64}\right)$ hekat

$= 90 + 10\left(\frac{8}{16} + \frac{1}{16} + \frac{8}{32} + \frac{1}{32} + \frac{8}{64} + \frac{1}{64}\right)$ hekat

$= 90 + 10\left(\frac{1}{2} + \frac{1}{4} + \frac{1}{8} + \frac{1}{16} + \frac{1}{32} + \frac{1}{64}\right)$ hekat $+ 50$ ro

$= 90 + 10\left(\frac{1}{2} + \frac{1}{4} + \frac{1}{8} + \frac{1}{16} + \frac{1}{32} + \frac{1}{64}\right)$ hekat $+ (10 \times 5)$ ro

$= 90 + 10\left[\left(\frac{1}{2} + \frac{1}{4} + \frac{1}{8} + \frac{1}{16} + \frac{1}{32} + \frac{1}{64}\right)$ hekat $+ 5$ ro$\right]$

$= 90 + 10(1 \text{ hekat}) = 90 + 10 = 100$ hekats.

Chapter 10

Geometry

The Greek historian Herodotus believed that geometry may have been a gift of the Nile. Aristotle argued that geometry rose in the Nile Valley because the priests had the leisure to develop theoretical knowledge. (Boyer, 1989, p. 24) Pythagorus had attainments in geometry corresponding with the ascertained fact that Egypt was the birthplace of the science (James, 1954, p. 43). This chapter contains some of the problems in geometry from the Hieratic and Demotic papyri, such as the Ahmes and Mosco Papyri. We will discuss some of the geometry problems of the Demotic papyri and show the differences between those and the problems in the Hieratic papyri. We will introduce the concept of the areas of rectangles, truncated triangles, circles, and hemispheres and the volumes of rectangular solids, cylinders, pyramids, and truncated pyramids. From the Demotic papyri, we will discuss some applications of the Pythagorean Theory and the area of the circle and circle segments. We will see the beginning of the Theory of Congruence and the Theory of Proportionality in the triangle and the idea of proof in geometry. Even though some of the methods were methods of estimation, it has scientific characteristics in its procedures, and the solution was remarkably close to the correct solution.

AREA OF RECTANGLES

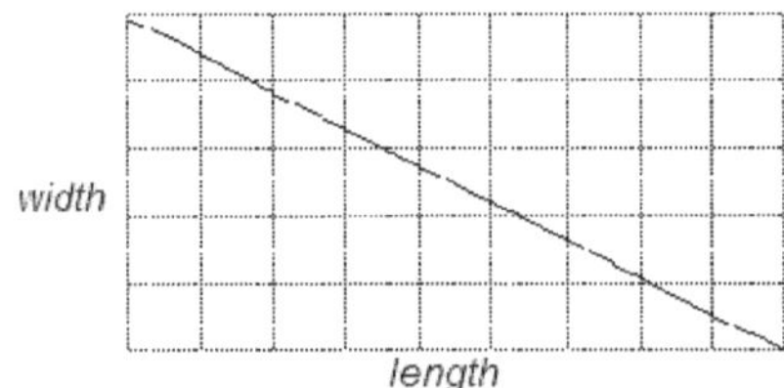

In all the problems in the papyri, the scribes calculated the area of the rectangle by multiplying the means of the two opposite sides. It may be that they arrived at the formula by counting the number of square units in the rectangle and found it to equal the length times the width.

Area of rectangles = length x width= the number of square units

Problem 49 (RMP)

The Ahmes Papyrus asked to find the area of a rectangle of land of 10 khet by one khat.

The scribe drew a rectangle:

Then he solved the problem by multiplying the length times the width:

1 khat = 100 cubits   10 khet = 1,000 cubits

Area = 100 x 1000 = 100,000 square cubits

$\quad$ =100,000 ÷ 100 = 1000 cubit strips

Note: The figure shows that the width is 2 khet; however, in the solution, the scribe considered the width as one khet.

AREA OF TRIANGLES

For the area of the triangle, the ancient Egyptians used the equivalent of the formula,

area $= \frac{1}{2}$ (base × height). In the words of the scribe, the area of the

triangle $= \frac{1}{2}$ teper × meret.

The teper translates to the base and the meret translates to either side or height. We will consider the meret to equal the height. In some problems, such as **Problem 51 (RMP),** the triangle is isosceles with a narrow base. Consider the meret equal to the side. In this case, the area will have an error by a small amount. The sketches of the triangle in the papyrus suggest that if a right-angle triangle is intended, then the one side and height are the same. The other side is the base.

Area of the triangle $= \frac{1}{2}$ base $\times$ height

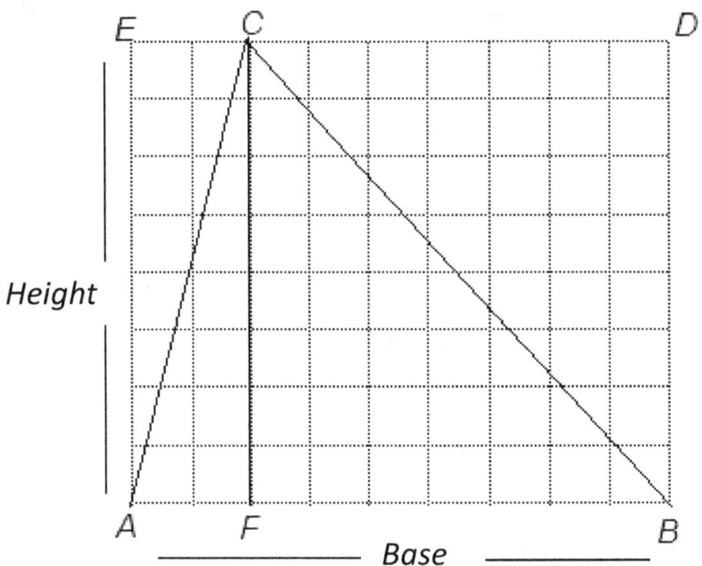

Maybe the scribe arrived at the formula by counting the number of square units of the triangle ABC, or by folding CB and AC. CB is the diagonal of the rectangle FBDC. The area of triangle FBC $= \frac{1}{2}$ the rectangle FBDC, and AC is the diagonal of rectangle AFCE; the area of triangle AFC $= \frac{1}{2}$ the rectangle AFCE.

Triangle FBC $+$ triangle AFC $= \frac{1}{2}$ rectangle FBDC $+ \frac{1}{2}$ rectangle AFCE

Triangle ABC $= \frac{1}{2}$ rectangle ABDE $= \frac{1}{2}$ (base) x (height) $= \frac{1}{2}\overline{AB} \times \overline{CF}$.

The scribe was aware of the relationship between a triangle and a rectangle with the same base and height. He always stated that when he found the area of the triangle that he was finding the area of the rectangle which equaled $\frac{1}{2}$ (base) (height).

Problem 51 (RMP)

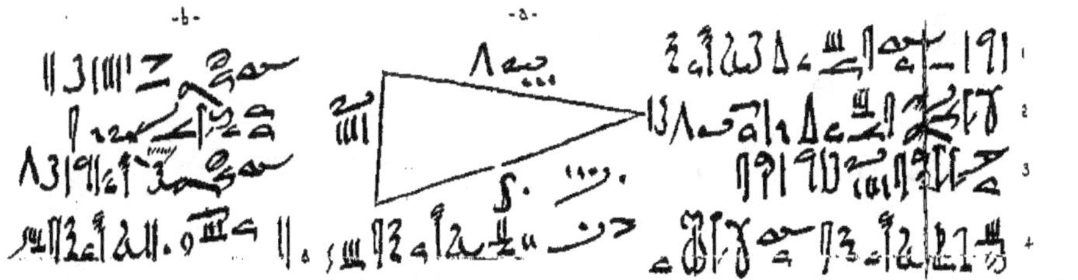

Translated as follows:

What is the area of the triangle of land if the side is 10 khet and the base is 4 khet?

The Solution

The scribe mentioned that in his solution, he was finding the area of the rectangle that equaled the area of the triangle.

$1 \rightarrow 400$;   4 khets = 400 cubits

$\frac{1}{2} \rightarrow 200$;   Multiply by $\frac{1}{2}$ to get the dimension of the rectangle.

$1 \rightarrow 1000$;   The other dimension of the equivalent triangle.

$2 \rightarrow 2000$;   The area of the rectangle in cubit strips.

He multiplied 2 by 1000 to get the area of the rectangle which equated with the area of the triangle. He thought the area was 2000 pairs of cubit strips. The scribe wrote the answer as 2 instead of 20 setats which equals 2 ten setats.

In the Mosco Papyrus, **Problem 4** (see page 268) is the same problem as above, which was to find the area of a triangle with a height of 10 khets and a base of 4 khets. The scribe took half of the 4 khets and multiplied by 10, and the answer was 20 square khets = 20 setats = 2 ten setats.

Note: If Ahmes means the side of the triangle and not the height, then:

$$\text{Height} = \sqrt{(10)^2 - (2)^2} = \sqrt{96} = 9.79.$$

$$\text{The area} = \frac{1}{2} \times \frac{400 \times 979}{10,000} = 19.58 \text{ setats.}$$

$$\text{The error} = 20 - 19.58 = .42 \text{ setat} = 2.15\%.$$

AREA OF CUT-OFF (TRUNCATED) TRIANGLE (TRAPEZOID)

For the area of the cut-off triangle, the ancient Egyptians used the equivalent of the formula $\frac{1}{2}$ (sum of the two bases) × height.

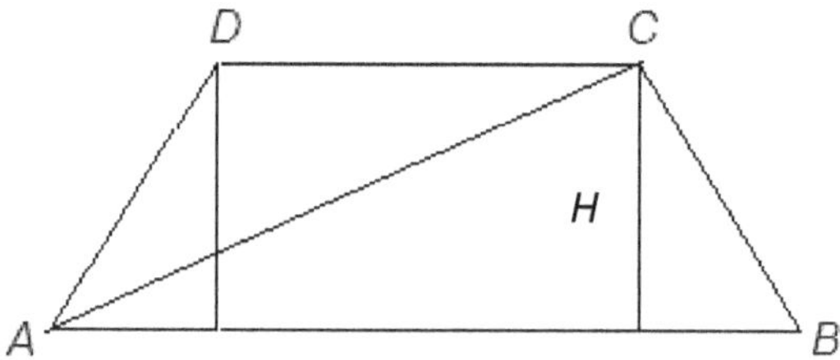

The scribe can arrive at this formula by finding the rectangle equivalent to the area of the trapezoid mentioned in the following problem by transferring the trapezoid to two triangles with the same height and with different bases.

$$\text{area} = \frac{1}{2} \text{ (sum of the two bases)} \times \text{height}$$
$$= \frac{1}{2} (\overline{AB} + \overline{DC}) \times H$$

However, the scribe used the incorrect formula which is the area equal to $\frac{1}{2}(\overline{DC} + \overline{AB}) \times \frac{1}{2}(\overline{AD} + \overline{CB})$, which is the product of the arithmetic mean of the opposite side. "There also seems to be evidence that the ancient Babylonians used this incorrect formula for the area of the quadrilateral" (Eves, 1969, p. 169).

Ahmes transferred the Isosceles triangle and trapezoid areas to the rectangles which equated the area. We see the beginning of the Theory of Congruence and the idea of the proven geometry.

Problem 52 (RMP)

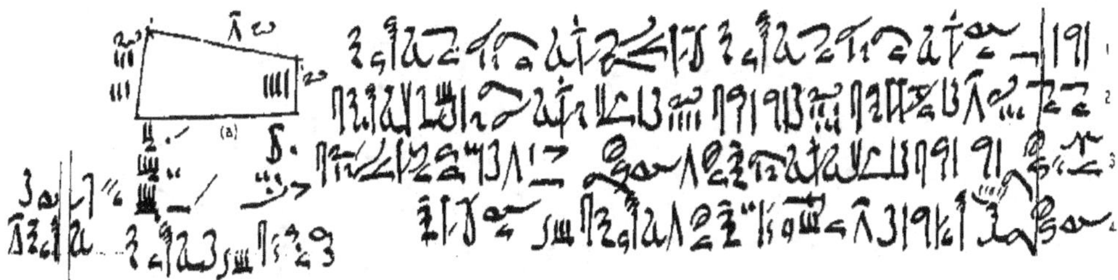

Translated as follows:

Find the area of the cut-off triangle with 20 khets for each side, 6 khets for the base, and 4 khets in its cut-off line.

The scribe drew the cut-off triangle as shown above, and it looks like the height = 20 khets.

The Solution

20 khets = 2000 cubits   6 khets = 600 cubits   4 khets = 400 cubits

The area $= \frac{1}{2}(600 + 400)(2000) \div 100 = 10{,}000$ cubit strips

$$= 10{,}000 \div 100 = 100 \text{ setats. } (10 \text{ ten setats.})$$

The correct solution is 99.78 setats.

The error is $\frac{100-99.78}{99.78} = .22\%$.

In the papyrus, 20 was written when 10 was shown in the above figure. The scribe also used the incorrect formula, Area $= \frac{(a+c)}{2} \times \frac{(b+d)}{2}$ for the area of the arbitrary quadrilateral with the successive side lengths a, b, c, and d. This formula does not give the right solution, but it gives a near-right solution in some special cases, like some problems above.

Problem 53 (RMP)

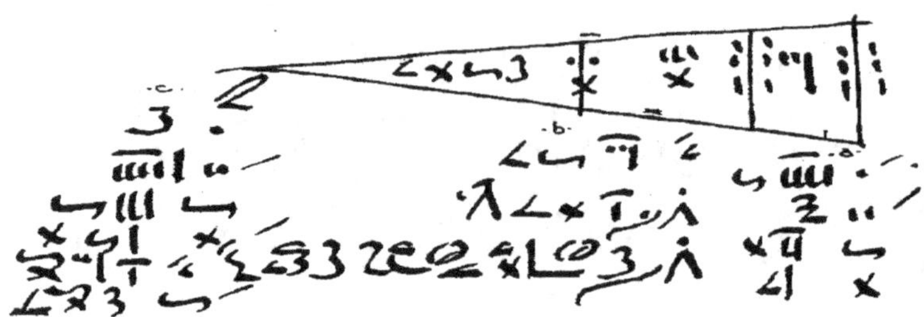

The scribe may be trying to get the area of the section of the Isosceles Triangle with a narrow base. The drawing suggests that 14 is the common length of the two equal sides and that the sections are made by the two lines parallel to the base. The scribe considered the height equal to one of the equal sides of the triangle. This does not make him correct, but nearly correct. If the Isosceles Triangle has a base of $4\frac{1}{2}$ cubits and each of the equal sides is 14 cubits, the area, according to Ahmes, would be $\frac{1}{2}\left(4\frac{1}{2}\right)(14) = 31\frac{1}{2}$ square cubits. However, the height of the triangle, using the Pythagorean Property, will be equal to 13.818, and the correct area of the triangle 31.09 square cubits. The Ahmes solution is a very close approximation to the correct solution.

He might apply an inaccurate rule for finding the area for the general quadrilaterals by taking the product of the arithmetic means of the opposite sides. In this case, if he applies half of the sum of the two sides by half of the third side to the area of the Isosceles Triangle, he would be using the zero concept as replacement for a magnitude in geometry.

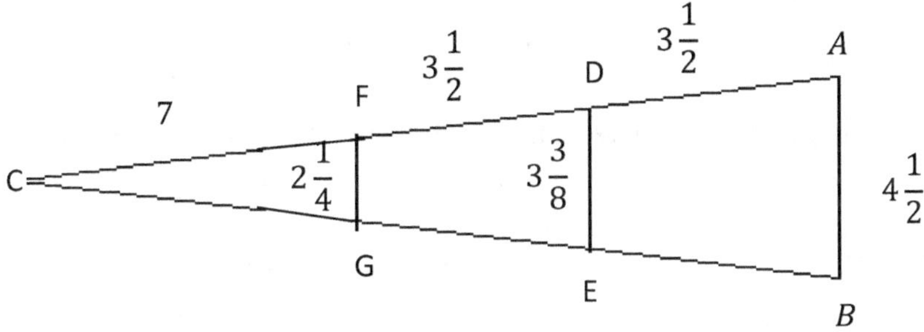

The Ahmes solution found two parts of the section of the triangle.

The area of triangle CGF $= \left[\frac{1}{2}\left(2\frac{1}{4}\right)\right] \times 7 = 7\frac{1}{2}\frac{1}{4}\frac{1}{8}$ setat.

The area of the section EBAD

The Solution

$$1 \longrightarrow 4\frac{1}{2}$$
$$2 \longrightarrow 9$$
$$\frac{1}{2} \longrightarrow 2\frac{1}{4}$$
$$\text{Total} \longrightarrow 15\frac{1}{2}\frac{1}{4}$$

$$\frac{1}{10}\left(15\frac{1}{2}\frac{1}{4}\right) = 1\frac{1}{2}\frac{1}{20}\frac{1}{40} \qquad 15\frac{1}{2}\frac{1}{4} - \frac{1}{10}\left(15\frac{1}{2}\frac{1}{4}\right) = 14\frac{1}{8}$$

$\frac{1}{10}$ of this is taken away. (*See note next page)

The shorter dividing line is $2\frac{1}{4}$ and this makes the base equal to $4\frac{1}{2}$. The longer dividing line is $3\frac{1}{4}\frac{1}{8}$.

The scribe put out 6 to be the length of both of the lines. If he wanted a rectangle equivalent to this area, he would take half of the sum of the bases of the trapezoid and multiply it by $3\frac{1}{2}$.

$$= \left(3\frac{1}{2}\frac{1}{4}\frac{1}{8}\frac{1}{16}\right)\left(3\frac{1}{2}\right) = 13\frac{1}{2}\frac{1}{4}\frac{1}{22} \text{ setats.}$$

But the scribe multiplied $4\frac{1}{2} \times 3 = 13\frac{1}{2}$ setats.

He did not put any answer but from his multiplication the answer will be $13\frac{1}{2}$ setats. But if he multiplied $3\frac{1}{2}$ by $4\frac{1}{2}$, the answer would be $15\frac{3}{4}$. He recognized this answer was too large because the top is less than the base in the trapezoid. The scribe then took $\frac{1}{10}$ of $15\frac{3}{4} = 1\frac{1}{2}\frac{1}{20}\frac{1}{40}$ * for that deficiency. And his answer is $14\frac{1}{8}$ setats, which is not correct (see previous page).

***Note: Sometimes the scribe took off a portion of his answer when he recognized the answer was greater than the correct area. This reminds us that the scribe took off $\frac{1}{9}$ of the diameter to determine the area of the circle.**

The answer makes sense if it is $13\frac{1}{2}$. This problem is difficult to explain, and the difficulty increases by numerical mistakes in the problem itself, or the solution of the scribe.

By using the Pythagorean Property, the height of the

$$\text{triangle} = \sqrt{(14^2) - \left(2\frac{1}{4}\right)^2} = 13\frac{1}{2}\frac{1}{5}\frac{1}{10}\frac{1}{50}$$

$$\frac{1}{2}\text{ height} = 6\frac{1}{2}\frac{1}{3}\frac{1}{15}\frac{1}{100}$$

$$\frac{1}{4}\text{ height} = \frac{1}{3}\frac{1}{15}\frac{1}{20}$$

The area of trapezoid $\text{ABED} = \frac{1}{2}\left(4\frac{1}{2} + 3\frac{1}{4}\frac{1}{8}\right) \times \left(3\frac{1}{3}\frac{1}{15}\frac{1}{20}\right) = 13\frac{1}{2}\frac{1}{5}\frac{1}{50}\frac{1}{100}$ setats, which is close to $13\frac{1}{2}$ setats.

The area of the small triangle $= \frac{1}{2}\left(2\frac{1}{4}\right)\left(6\frac{1}{3}\frac{1}{15}\frac{1}{100}\right) = 7\frac{1}{2}\frac{1}{5}\frac{1}{20}\frac{1}{50}$ setats which is very close to his answer.

The scribe was familiar with the theorem that **the segment that joins the midpoint of the two sides of the triangle is parallel to the third side and is half as long as the third side.**

Also, the scribe was familiar with the THEOREM OF PROPORTIONALITY which stated that: **In a triangle, if a line segment is drawn parallel to any one of the three sides to intersect the remaining two sides, they intersect at an identical ratio** (this theorem is known as **Thales Theorem.**)

THE PYTHAGOREAN PROPERTY

This property states that: **In the right-angle triangle, the square of the hypotenuse is equal to the squares of its two legs**.

Many of the mathematicians who wrote the history of mathematics, such as R. C. Archibald, B. L. van der Warren, Dirk J. Struik, etc., suggest that there is no evidence to say that the Egyptians were acquainted with the Pythagorean Property or a particular case of it. However, the Theory was known to the Egyptian priests. When he came to Egypt, Pythagoras sacrificed to the Muses when the Egyptian priests explained to him the properties of the right-angle triangle. (Philarch de Repugn. Stoic 2 p. 1089; Demetrius; Antisthenes; Cicero de Natura Decorum III, 36) Taken from James, 1954, p. 43.

Problems 24-31 in the Cairo Papyrus are concerned with a pole of varying heights. In **Problems 24-26**, the pole is moved outward a certain distance with a resultant lowering of the top, and the amount of such a lowering is then determined. These are right-angle triangle problems with the pole becoming the hypotenuse. The base is given and the height is determined by the Pythagorean Property as shown in **Problems 24 and 28** from the Cairo Papyrus.

Problem 24 (Cairo Papyrus)

A pole is 10 cubits when erect and its foot is moved outward 6 cubits. By what amount is the top of the pole lowered?

$a = 10;$ $y = 6;$ $x =$ the number of cubits the top is lowered.

The Solution

$a =$ length of the pole; $y =$ the distance the foot is moved outward.

$a^2 = 10 \times 10 = 100$

$y^2 = 6 \times 6 = 36$

$100 - 36 = 64$ $a^2 - y^2 = (a - x)^2$

$\sqrt{64} = 8$ $a - x = \sqrt{(a^2 - y^2)}$

$10 - 8 = 2$ $x = a - \sqrt{(a^2 - y^2)}$

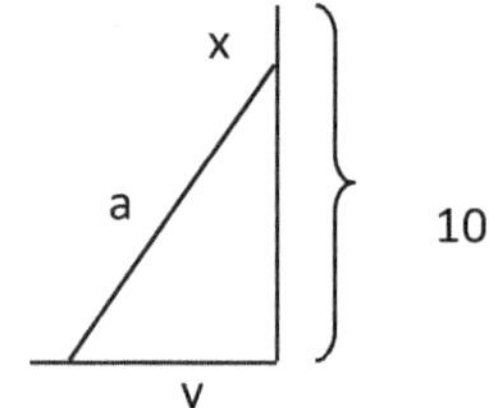

$x =$ the amount the top of the pole is lowered (2 cubits).

Problem 28 (Cairo Papyrus)

The erect pole is $14\frac{1}{2}$ cubits high. How far is the foot of the pole moved outward if the top of it is moved 4 cubits downward?

$a = 14\frac{1}{2};$ $x = 4;$ $y =$ the distance the foot is moved outward.

The Solution

$14\frac{1}{2} \times 14\frac{1}{2} = 210\frac{1}{4} = a^2$

$14\frac{1}{2} - 4 = 10\frac{1}{2} = a - x$

$10\frac{1}{2} \times 10\frac{1}{2} = 110\frac{1}{4} = (a - x)^2$

$210\frac{1}{4} - 110\frac{1}{4} = 100 = a^2 - (a - x)^2$

Reducing 100 to its square root is 10; $\sqrt{a^2 - (a - x)^2} = y$

The base of the pole was moved outward 10 cubits.

Problem 30 (Cairo Papyrus)

Given that the top of an erect pole has been lowered 2 cubits, and the foot of the pole has been moved outward 6 cubits, determine the original height of the pole.

The Solution

$6 \times 6 = 36 = L^2$

$2 \times 2 = 4 = k^2$

$4 + 36 = 40 = k^2 + L^2$

$2 \times 2 = 4 = 2k$

$40 \div 4 = 10 = \dfrac{k^2 + L^2}{2k} = x$

The original height of the pole is $x = 10$ cubits.

Problems 34 and 35 (Cairo Papyrus)

The area of the rectangle and the diagonal are given and we are asked to determine the side. The scribe uses the Pythagorean Property to solve these problems.

Problem 34 (Cairo Papyrus)

Given a rectangle with the area $(xy) = 60$ square cubits, with a diagonal of $d = 13$, determine the two sides.

$xy = 60 \qquad d = 13 \qquad d^2 = x^2 + y^2 = 169$

The Solution

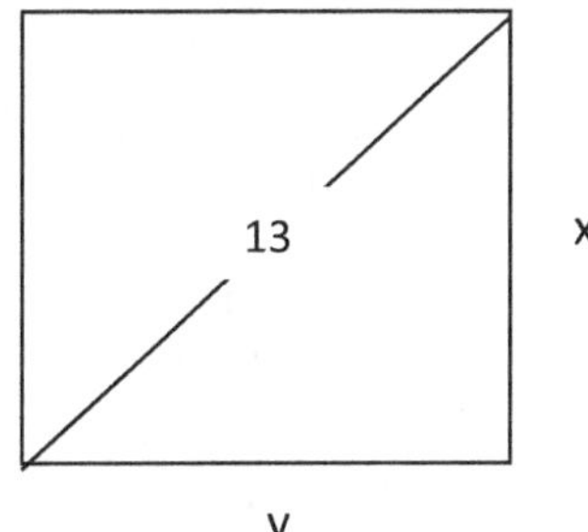

$13 \times 13 = 169 = d^2$

$60 \times 2 = 120 = 2a$

$169 + 120 = 289 = d^2 + 2a$

Reduce it to its square root and the result is $17 = \sqrt{d^2 + 2a}$

$169 - 120 = 49 = d^2 - 2a$

Reduce it to its square root and the result is $7 = \sqrt{d^2 - 2a}$

$$17 + 7 = 24 = \sqrt{d^2 + 2a} + \sqrt{d^2 - 2a}$$
$$\tfrac{1}{2} \text{ of } 24 = 12 = \text{height} = \tfrac{1}{2}\left(\sqrt{d^2 + 2a} + \sqrt{d^2 - 2a}\right) \text{ cubits}$$
$$17 - 7 = 10 = \sqrt{d^2 + 2a} - \sqrt{d^2 - 2a}$$
$$\tfrac{1}{2} \text{ of } 10 = 5 = \text{width} = \tfrac{1}{2}\left(\sqrt{d^2 + 2a} - \sqrt{d^2 - 2a}\right) \text{ cubits.}$$

The answer is that the land is 12 cubits by 5 cubits in size.

<u>The Proof</u>

$$12 \times 12 = 144$$
$$5 \times 5 = 25$$
$$144 + 25 = 169$$

Reduce 169 to its square root of 13 cubits, which is equal to the diagonal.

THE AREA OF A CIRCLE (RMP)

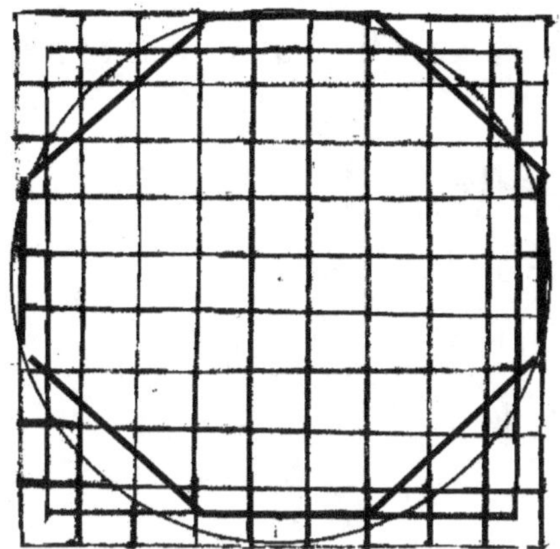 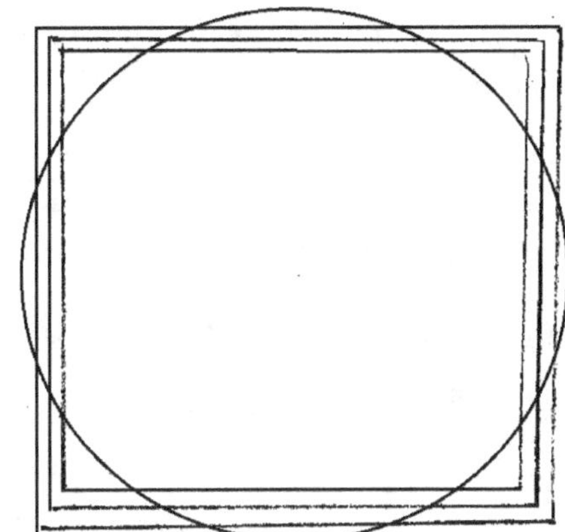

The scribe determined the area of the circle by subtracting $\frac{1}{9}$ of the diameter and squared the result. He determined the area of the circle to be the area of circumscribing squares with a side equal to $\frac{8}{9}$ of the diameter. If the diameter of the circle is 'd', then the area of the circle will be

$$\left(d - \frac{d}{9}\right)\left(d - \frac{d}{9}\right) = \left(d - \frac{d}{9}\right)^2 = \left(\frac{8}{9}d\right)^2 \; ; \quad d = 2r \quad \text{'r' is the radius}$$

Problem 48 (RMP)

The problem shows us how Ahmes arrived at this result by forming an octagon from the square of the side of nine units by trisecting the sides and cutting off the four corners of the Isosceles Triangles. I suggest the following procedure to arrive at his result. The area of the octagon will not differ greatly from the area of the circle inscribed within the square. The area of the octagon = the area of the square with side 'd' less than the area of four triangles with a height of $\frac{d}{3}$ and the base $\frac{d}{3}$. If the side of the square is 'd',

and the area of the square = d^2, the area of the 4 triangles = $4\left[\left(\frac{1}{2}\right)\left(\frac{d}{3}\right)\right]\frac{d}{3}$

$$= \frac{2}{9}d^2.$$

The area of the octagon $= d^2 - \frac{2}{9}d^2 = d^2 - \frac{18}{81}d^2 = \frac{63}{81}d^2 \approx \frac{64}{81}d^2$

$= \left(\frac{8}{9}d\right)^2$, which the scribe felt is the area of the circle. We know that the area of the circle $= \pi r^2$.

$$\pi r^2 \approx \left(\frac{8}{9}d\right)^2 = \left(\frac{8}{9}\right)^2 (4r^2) = 3.16049r^2; \quad \pi = \left(\frac{8}{9}\right)^2 \times 4$$

$= \pi \approx 3.160493827 \qquad \text{The correct } \pi \approx 3.141592654$

$\text{Error} = 3.16049382 - 3.141592654 = .0189011736$

$\approx .6\%$ which is a remarkably close approximation.

Problem 48 (RMP)

The octagon which is
inscribed in the
square to the right is
the approximation of
the circle.

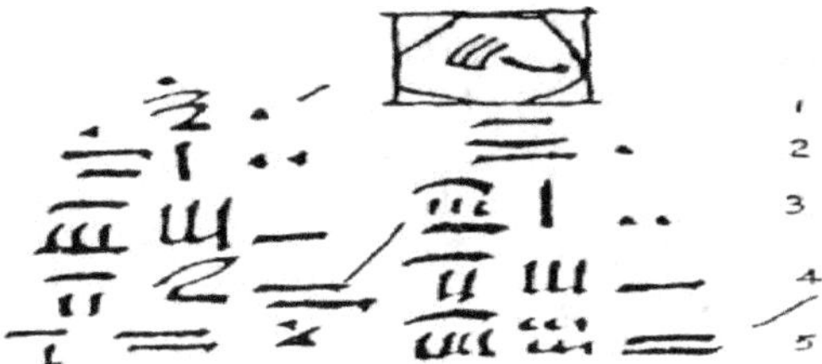

Here is the first time to compare the curved and straight-line configuration by circumscribing the square over the circle and finding the area of the polygon in the square to be the approximation of the area of the circle.

Translated as follows:

Compare the area of the circle and its circumscribing square.

<u>The Solution</u>

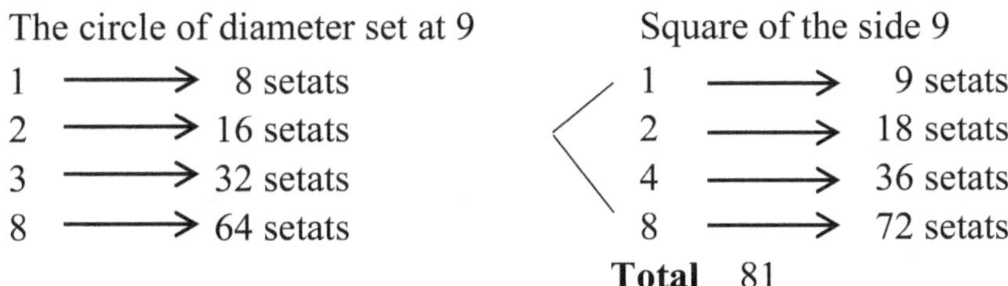

The circle of diameter set at 9   Square of the side 9

1	⟶	8 setats
2	⟶	16 setats
3	⟶	32 setats
8	⟶	64 setats

1	⟶	9 setats
2	⟶	18 setats
4	⟶	36 setats
8	⟶	72 setats

Total 81

π is the ratio of the circumference of any circle to its diameter. It was given by Archimedes to be less than $3\frac{1}{7}$, but not greater than $3\frac{10}{71}$. He used the inscribed with 96 sides. In 1579, Francois Viete used a polygon having $6(2^{16})$ sides to find π, correct to 9 decimal places. Luddph van Cemlen of Germany used polygons having 2^{64} sides to compute π to 35 decimal places (Smith, 1958, p. 310).

Problem 50 (RMP)

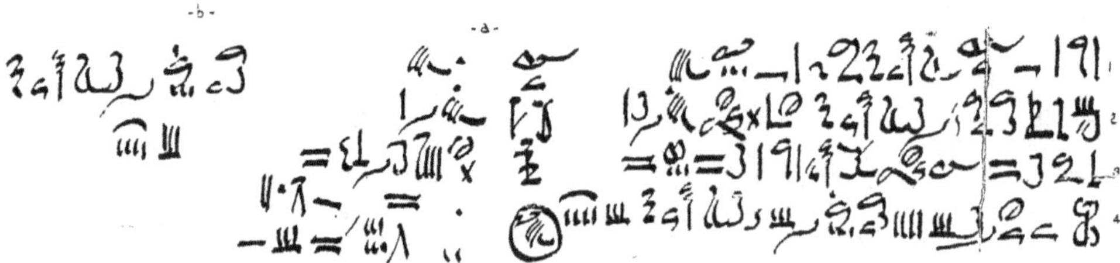

Translated as follows:

A round field has a diameter of 9 khets. What is the area?

To solve the problem, the scribe took $\frac{1}{9}$ of the diameter and subtracted it from the diameter of 9.

$9 - 1 = 8$;   square 8 and the result is 64 setats.

The real area $= \pi r^2 = \pi(4.5)^2 = 63.617$ setats.

THE AREA OF THE CIRCLE IN THE CAIRO PAPYRUS

The scribe noted the relationship between the diameter, 'd', and the area of the circle 'A' as follows:

$$d = \sqrt{A + \frac{A}{3}} \Rightarrow d = \sqrt{\frac{4A}{3}} \Rightarrow d^2 = \frac{4}{3}A$$

$$\Rightarrow A = \frac{3}{4}d^2 \Rightarrow A = \frac{3}{4}(2r)^2 = 3r^2 \qquad d = 2r$$

$$\therefore \pi = 3$$

If $\pi = 3$, then the circumference 'c' of the circle is

$$c = 2\pi r = \pi(2r) = \pi d = 3d \text{ and } d = \frac{c}{3};$$

$$\therefore A = \frac{3}{4}d^2 = \frac{3d}{4} \times \frac{3d}{3} = \frac{c}{4} \times \frac{c}{3} = d^2 - \frac{d^2}{4}.$$

This is the formula for the area of the circle.

The scribe used all of these relationships to solve **Problems 32, 33, 36, 37, and 38** in the Cairo Papyrus.

Problems 32 and 34 (Cairo Papyrus)

The area of the circle was given, and the problem asks for the diameter.

<u>The Solution</u>

To solve this problem, the scribe used the formula $d = \sqrt{A + \frac{A}{3}}$

where d = diameter, and A = the area of the circle.

<u>The Proof</u>

The scribe found c = 3d and shows that $A = \frac{c}{4} \times \frac{c}{3}$.

Problem 32 (Cairo Papyrus)

Area of the circle 'A' = 100 square cubits. Find the diameter 'd'.

Then $A + \frac{A}{3} = 133\frac{1}{3}$. The square root of $133\frac{1}{3} = 11\frac{1}{2}\frac{1}{120} = d$.

The Proof

$$c = 3d = 3\left(11\frac{1}{2}\frac{1}{120}\right) = 34\frac{1}{2}\frac{1}{10}\frac{1}{20} \; ; \; \frac{1}{4}c = \left(8\,\frac{1}{2}\frac{1}{10}\frac{1}{20}\frac{1}{120}\right);$$

$$A = \frac{c}{3} \times \frac{c}{4} = \left(11\frac{1}{2}\frac{1}{120}\right)\left(8\,\frac{1}{2}\frac{1}{10}\frac{1}{20}\frac{1}{120}\right) = 100 \text{ square cubits.}$$

The correct answer is $100\frac{3}{800} = 100\,\frac{1}{400}\frac{1}{800}$.

In a problem written on a clay tablet by the Babylonians (probably 1800-1600 BC), the area of the circle was expressed sexagesimally as

$A = (0;5)C^2$. In our decimal system, $A = \frac{5}{60}C^2$ or $\frac{1}{12}C^2$, which makes π equal to 3. There is also a statement in the Bible in I Kings 7:30 reading that the basin built in connection with King Solomon's temple in 1000 BC was 10 cubits from one brim to the other, and a line of 30 cubits did encompass it around, which made $\pi = 3$ (Von Baravalle, 1969, p. 149).

THE AREA OF THE SEGMENT OF A CIRCLE

The ancient Egyptian mathematicians developed a formula to find the area of the circular segment of a circle.

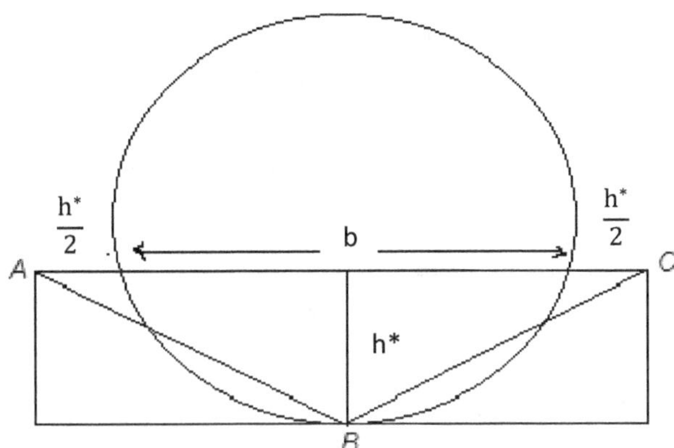

The area of the segment a $\approx \Delta \approx$ ABC $= \frac{(h^*+b)}{2} \times h^*$

where h^* = height and b = base of the circular segment.

If $\pi = 3$, the area of the triangle ABC will be a close approximation to the area of the circular segment.

If $h^* = \frac{d}{2}$ then b = d, and a $= \frac{1}{2}\left(\frac{d}{2} + d\right)\frac{d}{2} = \frac{1}{2}\left(3\frac{d^2}{4}\right).$

If $\pi = 3$ and d = 2r, then a $= \frac{1}{2}\pi r^2$ = half of the area of the circle.

Problem 36 (Cairo Papyrus)

We are given a circular plot of land which is laid out in an equilateral triangle with a given side. We are asked for the area of the plot.

The Solution

The scribe found the area of the triangle and found the area of one of the circular segments using the formula a $= \frac{(h^*+b)}{2} \times h^*$.

For 3 circular segments, 3a $= \frac{3}{2}$ (h* + b) h* + the area of the triangle equals the plot area.

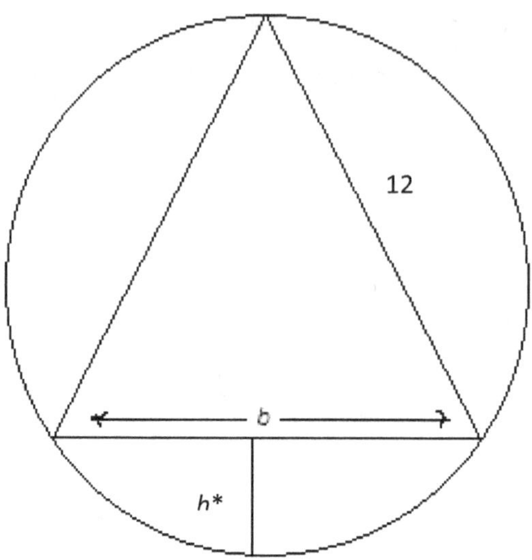

Problem 36 (Cairo Papyrus)

Given a circular plot of land in which is laid out an equilateral triangle with sides of 12 divine cubits each and points touching the circle, determine the area of the plot.

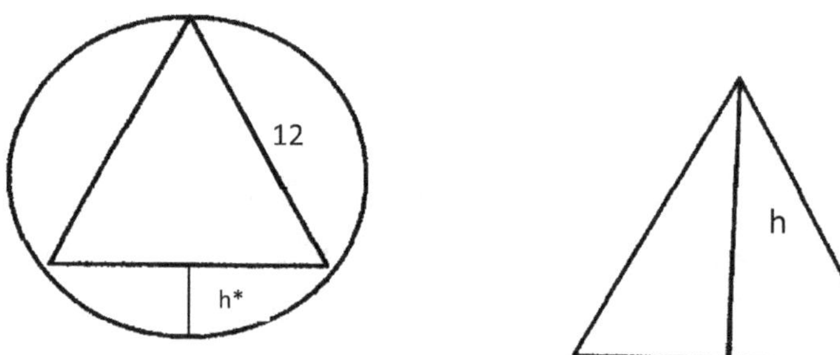

The scribe found the height of the triangle and multiplied it by half of the base $a = \frac{1}{2}bh$; then he found the area of the circular segment using the formula $a = \left(\frac{h^* + b}{2}\right)h^*$, and he multiplied the area of the circular segment by

3. He added the total area of the 3 circular segments to the area of the triangle to find the area of the circle. To prove his work, he found the diameter of the circle, then found the area of the circle using the formula $\frac{c}{3} \times \frac{c}{4}$ by finding the circumference, then he proved the area of four pieces = the area of the circle.

The Solution

We will consider four pieces of land, the triangle and the three circular segments. We will get the area of the triangle.

Multiply 12 x 12; the result is 144.

Multiply 6 x 6; the result is 36.

Subtract 36 from 144; the result is 108.

Find the square root of 108; the result is $10\,\frac{1}{3}\,\frac{1}{20}\,\frac{1}{120}$.

This is the middle height of the triangle.

The number of the base is 12, half of it is 6.

The area of the triangle, multiply 6 by $10\,\frac{1}{3}\,\frac{1}{20}\,\frac{1}{120} = 62\,\frac{1}{3}\,\frac{1}{60}$ square cubits.

We shall take $\frac{1}{3}$ of $10\,\frac{1}{3}\,\frac{1}{20}\,\frac{1}{120}$, the result is $3\,\frac{1}{3}\,\frac{1}{10}\,\frac{1}{60}\,\frac{1}{120}\,\frac{1}{180} = h^*$.

This is the height of the circular segment.

$h^* = 3\,\frac{1}{3}\,\frac{1}{10}\,\frac{1}{60}\,\frac{1}{120}\,\frac{1}{180}$

We shall add $3\,\frac{1}{3}\,\frac{1}{10}\,\frac{1}{60}\,\frac{1}{120}\,\frac{1}{180}$ to 12 cubits.

The result is $15\,\frac{1}{3}\,\frac{1}{10}\,\frac{1}{60}\,\frac{1}{120}\,\frac{1}{180}$ cubits (h* + b).

Then we will take half of it; the result is $7\,\frac{2}{3}\,\frac{1}{20}\,\frac{1}{120}\,\frac{1}{240}\,\frac{1}{360}\left(\frac{h^{*}+b}{2}\right)$ cubits.

Then we will multiply this by $3\,\frac{1}{3}\,\frac{1}{10}\,\frac{1}{60}\,\frac{1}{120}\,\frac{1}{180}$; the result is $26\,\frac{5}{6}\,\frac{1}{10}\left(\frac{h^{*}+b}{2}\right)h^{*}$ square cubits.

Then we will multiply 3 times; the result is $80\,\frac{2}{3}\,\frac{1}{10}\,\frac{1}{30}$ square cubits.

Add this to $62\,\frac{1}{3}\,\frac{1}{60}$; the result is $143\,\frac{1}{10}\,\frac{1}{20}$ square cubits.

This is the area of the circular plot.

The Proof

The height of the triangle is $10\,\frac{1}{3}\,\frac{1}{20}\,\frac{1}{120}$ cubits.

We add $\frac{1}{3}$ of the height, $3\,\frac{1}{3}\,\frac{1}{10}\,\frac{1}{60}\,\frac{1}{120}\,\frac{1}{180}$; the result is $13\,\frac{5}{6}\,\frac{1}{45}$ cubits.

We multiply $13\,\frac{5}{6}\,\frac{1}{45}$ by 3 to get 'C', the circumference.

The result is $41\,\frac{1}{2}\,\frac{1}{15}$ cubits.

$\frac{1}{3}$ of the circumference $13\,\frac{5}{6}\,\frac{1}{45}$ and $\frac{1}{4}$ of the circumference is $10\,\frac{1}{3}\,\frac{1}{20}\,\frac{1}{120}$.

Multiply $\frac{c}{3}\times\frac{c}{4}=13\,\frac{5}{6}\,\frac{1}{45}$ by $10\,\frac{1}{3}\,\frac{1}{20}\,\frac{1}{120}$.

The result is $143\,\frac{5}{6}\,\frac{1}{10}\,\frac{1}{30}$ square cubits.

The difference is $\frac{2}{3}\,\frac{1}{10}\,\frac{1}{20}$ square cubits.

To solve the problem in a modern approach,

The area of the triangle $=\frac{1}{2}bh$.

$b=12\qquad h=\sqrt{144-36}=\sqrt{108}=10.39304$ cubits.

The area of the triangle $=\frac{1}{2}(12)(10.39304)=62.35824$ square cubits.

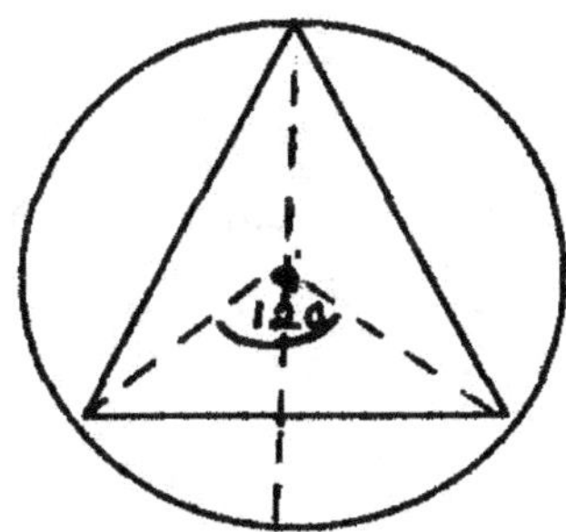

The area of the segment a $= \frac{1}{2}r^2(\theta - \sin\theta)$ where 'r' is the radius

and 'θ' is the angle in radians subtended, and the center of the circle by the

arc, r $= \frac{1}{2}\left(h + \frac{1}{3}h\right) = \frac{1}{2}(10.39304) + \frac{1}{3}(10.39304) = 6.928$ cubits.

The area of the circular segment $\frac{1}{2}r^2(\theta - \sin\theta)$ $\theta = \frac{2\pi}{3}$ radius

$$\theta = 120 \text{ degrees}$$

$\frac{1}{2}(6.928)^2\left[\frac{2\pi}{3} - \sin 120\right] = 29.47914.$

The area of the three circular segments is 3 x 29.4791 = 88.4374 square

    cubits.

The total area of the four pieces is 62.3582 + 88.4374 = 150.7956 square

    cubits.

The Proof

The radius of the circle is 6.928.

Area of the circle $= \pi r^2 = \pi(6.928)^2$

$$= 150.7876 \text{ square cubits.}$$

If we take $\pi = 3$,

then the area $= 3r^2 = 3(6.928)^2 = 143.99$ square cubits.

Problem 37 (Cairo Papyrus)

A square was laid in a circular plot. To solve this problem, the scribe found the area of the square and the four circular segments and added them to find the area of the circular plot. See **Problem 37** in the Cairo Papyrus.

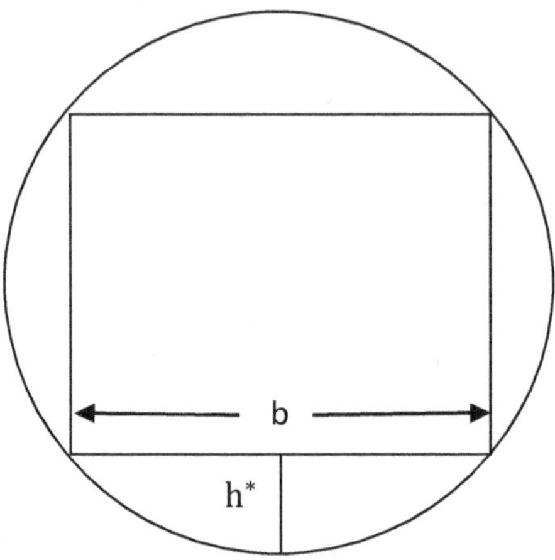

THE AREA OF A HEMISPHERE AND THE AREA OF A SEMI-CYLINDER

Problem 10 (Mosco Papyrus)
This problem received interest among present-day mathematicians, as we will explain.

We are given a basket with a mouth of $4\frac{1}{2}$ cubits. What is the surface area? To solve this, the scribe doubled the diameter of $4\frac{1}{2}$ to equal 9.

Calculate $\frac{1}{9}$ of 9 because the basket is half an egg.

The result is equal to 1. Take 1 from 9; the remainder is 8.

Calculate $\frac{1}{9}$ of 8. The result is $\frac{2}{3}\frac{1}{6}\frac{1}{18}$. Take $\frac{2}{3}\frac{1}{6}\frac{1}{18}$ from 8; remainder is $7\frac{1}{9}$.

$7\frac{1}{9} \times 4\frac{1}{2} = 32$ square cubits. This is the correct area.

***Note**:* We will later explain the different solutions to this problem.

To solve this problem, Richard Gillings first considered the basket as a hemisphere and T. Peet as a semi-cylinder.

If we consider the basket as a hemisphere, then the scribe was familiar with the area of the surface of the area of a hemisphere. The area 'A' of the basket would

$$= \frac{1}{2}(4\pi r^2) = 2\pi r^2.$$

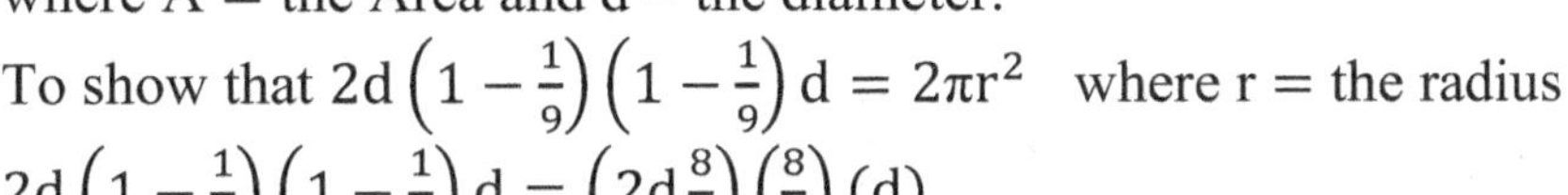

The scribe used the formula:

$$A = 2d\left(1 - \frac{1}{9}\right)\left(1 - \frac{1}{9}\right)d \qquad d = 2r$$

where A = the Area and d = the diameter.

To show that $2d\left(1 - \frac{1}{9}\right)\left(1 - \frac{1}{9}\right)d = 2\pi r^2$ where r = the radius

$$2d\left(1 - \frac{1}{9}\right)\left(1 - \frac{1}{9}\right)d = \left(2d\frac{8}{9}\right)\left(\frac{8}{9}\right)(d).$$

$$2d\left(\frac{8}{9}\right)\left(\frac{8}{9}\right)d = 2(2r)\left(\frac{8}{9} \times \frac{8}{9}\right)2r \quad \text{where } d = 2r$$

$$= 4r\left(\frac{8}{9} \times \frac{8}{9}\right)2r = 2r\left(\frac{8 \times 8 \times 4}{81}\right)r$$

$$= 2\left(\frac{256}{81}\right)r^2 = 2\pi r^2 \text{where } \pi = \frac{256}{81}$$

$$= \frac{1}{2} \text{ the area of the sphere.}$$

The summary of the scribe's solution:

Double the diameter: $2d = 2 \times 4\frac{1}{2} = 9.$

Calculate $\frac{1}{9}$ of 9; the result = 1 cubit.

Take 1 from 9; the result = 8 cubits. $\qquad\qquad \left(2d - \frac{2d}{9}\right).$

Calculate $\frac{1}{9}$ of 8; the result $= \frac{1}{3}\frac{1}{6}\frac{1}{18}$ cubits. $\qquad \frac{1}{9}\left(2d - \frac{2d}{9}\right).$

Take $\frac{1}{3}\frac{1}{6}\frac{1}{18}$ from 8; the result is $7\frac{1}{9}$ cubits. $\quad = 2d\left(1 - \frac{1}{9}\right) - \frac{2d}{9}\left(1 - \frac{1}{9}\right).$

$$= 2d\left(1 - \frac{1}{9}\right)\left(1 - \frac{1}{9}\right).$$

Multiply $7\frac{1}{9} \times 4\frac{1}{2} = 32$ square cubits. $\qquad = 2d\left(1 - \frac{1}{9}\right)\left(1 - \frac{1}{9}\right)d .$

The result is 32 square cubits; this is the area $= 2\pi r^2$

T. Peet suggests another approach to solving this problem. Considering this as a semi-cylinder with

$A = \frac{1}{2}\pi dh$ where $\pi = \frac{256}{81}$ and $d = 4\frac{1}{2}$ and $h = 4\frac{1}{2}$ cubits;

$A = \frac{1}{2}\left(\frac{256}{81}\right)\left(4\frac{1}{2}\right)\left(4\frac{1}{2}\right) = 32$ square cubits.

$A = \frac{1}{2}\pi dh$, the formula for the area of a semi- cylinder.

The result is $A = \frac{1}{2}\left(\frac{256}{81}\right)\left(4\frac{1}{2}\right)\left(4\frac{1}{2}\right) = 32$ square cubits, which is the right answer.

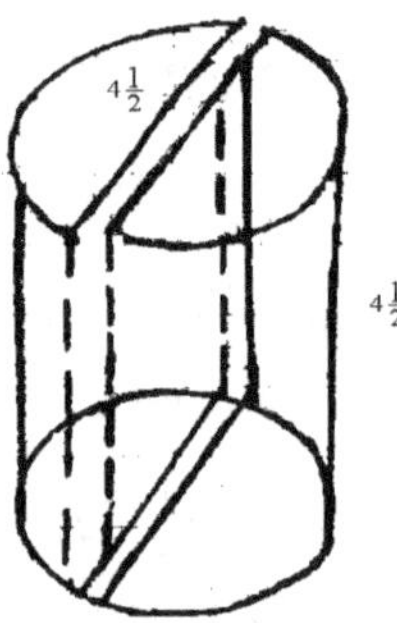

Interpretation of the translation of the problem:

Peet suggests the basket is a semi-cylinder with a diameter of $4\frac{1}{2}$ cubits and a height of $4\frac{1}{2}$ cubits. He adds two more terms, again the diameter of $4\frac{1}{2}$ cubits and a height of $4\frac{1}{2}$ cubits. This was done because $4\frac{1}{2}$ was written twice in the solution of the problem. This gives the impression that the problem was asking for the area of a Quonset hut type of roof.

The area $= \frac{1}{2}\pi dh$; therefore, the solution is $\frac{1}{2}\left(\frac{256}{81}\right)\left(4\frac{1}{2}\right)\left(4\frac{1}{2}\right) = 32$ square cubits, giving the same answer.

If Peet's interpretation is right, then the ancient Egyptian knew that the length of the circumference of a circle is equal to πd and, therefore, the length of the circumference of the semi-circle $= \pi r$.

The length of the circumference $\frac{1}{2}\left[4\left(1-\frac{1}{9}\right)\left(1-\frac{1}{9}\right)d\right] = 2\pi r$

Where $\pi = 4\left(1-\frac{1}{9}\right)\left(1-\frac{1}{9}\right) = \frac{256}{81}$

C = the circumference $= 4\left(1-\frac{1}{9}\right)\left(1-\frac{1}{9}\right)d$

The length of the semi-circle $= 2\left(1-\frac{1}{9}\right)\left(1-\frac{1}{9}\right)d$

<u>The Proof</u>

$$= 4\left(1-\frac{1}{9}\right)\left(1-\frac{1}{9}\right)d$$

$$= 4\left(\frac{8}{9}\right)\left(\frac{8}{9}\right)2r = 2\pi r$$

<u>The Expected Solution</u>

The area of the semi-cylinder $= \left(\frac{1}{2}C\right)h$ (where 'C' is the circumference).

$$= \frac{1}{2}\left[4\left(1-\frac{1}{9}\right)\left(1-\frac{1}{9}\right)d\right]h \qquad d = 4\frac{1}{2} \quad \text{and} \quad h = 4\frac{1}{2}$$

$$= \frac{1}{2}\left(\frac{256}{81}\right)\left(4\frac{1}{2}\right)\left(4\frac{1}{2}\right) = 32 \text{ square cubits.}$$

Note: If the ancient Egyptian considered this as a hemispherical surface, "such a result antedated the oldest known calculation of a hemispherical surface by some 1500 years. This would be almost too amazing to be true." (Boyer & Merzbach, 1989, p. 24)

VOLUME AND CAPACITY

The ancient Egyptian mathematicians were able to determine the volume of prisms, cylinders, and pyramids.

Problems 41-46 (RMP)

The author calculates the amount of grain in a bin of given dimensions.

Problem 44 (RMP)

This is an example of finding the volume of a rectangular granary by multiplying the three dimensions together, or the area of the base multiplied by the height.

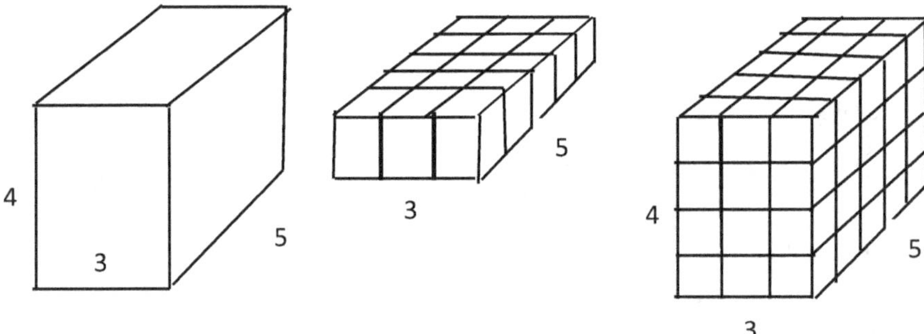

The ancient mathematician may have found the volume as we find it now.

We have a rectangular granary with a base of 3 x 5 units, and its height is 4 units. If we cover the base with the unit cubic cubits blocks, which take 15 blocks to cover the rectangle, we need four of these to do so. We can imagine the prism as building stacks of cubes on top of each other until the prism is completely filled. Since there are four levels to the prism, it will be completely filled by 3 x 5 x 4 = 60 cubic cubits.

Problem 44 (RMP)

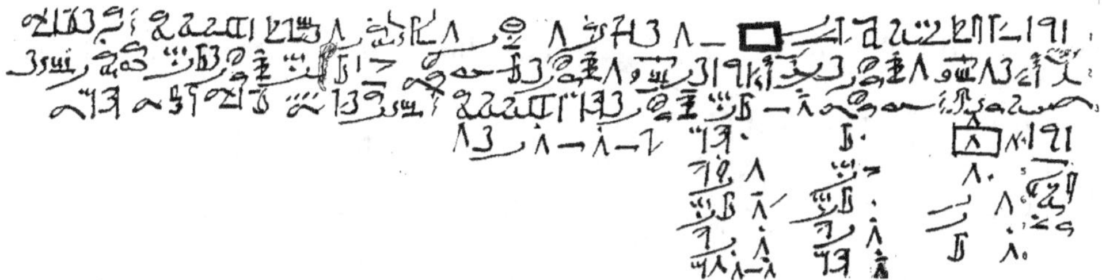

Translated as follows:

Given a rectangular granary with a length of 10 cubits and a width of 10 cubits and a height of 10 cubits, what is the amount of grain that goes into it?

The Solution

The scribe multiplied the width by the length by the height.

The amount of grain $= 10 \times 10 \times 10 = 1000$ cubed cubits.

$$= 1000 + \frac{1}{2}(1000) = 1500 \text{ khar.}$$

$$= 1500 \times \frac{1}{20} = 75 \text{ hundred quadruple hekats.}$$

20 khar = 100 quadruple hekats.

Problem 45 is the reverse of Problem 44 in the RMP. Given the value of the volume of a rectangular granary into which there have gone 7500 quadruple hekats of grain, find the dimensions.

The Solution

Multiply 75 x 20. It takes 1500 khar. Take $\frac{1}{10}$ of 1500 = 150.

Then take $\frac{1}{10} \times \frac{1}{10}$ of 1500 = 15. Take $\frac{2}{3}\left(\frac{1}{10} \times \frac{1}{10} \text{ of } 1500\right) = 10$ cubits.

He assumed that the rectangular granary had a width of 10 cubits and a length of 10 cubits. (Instead of taking $\frac{2}{3}$ at the beginning to reduce the contents to cubed cubits, as he would have done if he had exactly reversed the process of the preceding solution, the author takes $\frac{2}{3}$ of the last quotient to find the third dimension.)

VOLUME OF CYLINDERS

The Egyptian mathematician sees the cylinder with a diameter of the base equal to 'd', the same as a rectangular solid with a square base where each side is $\left(d - \frac{1}{9}d\right)$ and height = h.

The volume is $V = \left(d - \frac{1}{9}d\right)\left(d - \frac{1}{9}d\right)h.$

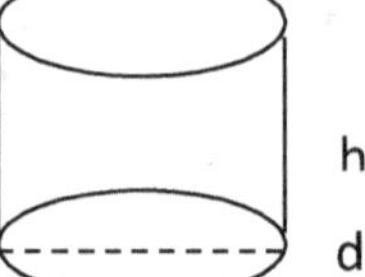

If you want the result by the khar, which is the cubed cubit = $1\frac{1}{2}$ khar, then

$$V = \left(d - \frac{1}{9}d\right)\left(d - \frac{1}{9}d\right)h\left(1\frac{1}{2}\right)\text{khar.}$$

In the Kahun Papyrus, the scribe used the equivalent formula,

$$V = \left(d + \frac{1}{3}d\right)\left(d + \frac{1}{3}d\right) \times \frac{2}{3}h.$$

The Proof

$$\left(d - \frac{1}{9}d\right)\left(d - \frac{1}{9}d\right) \times \frac{3}{2}h = \left(d + \frac{1}{3}d\right)\left(d + \frac{1}{3}d\right)\frac{2}{3}h.$$

$$V = \left(d - \frac{1}{9}d\right)\left(d - \frac{1}{9}d\right)\frac{3}{2}h.$$

$$= \frac{8}{9}d \times \frac{8}{9}d \times \frac{3}{2}h.$$

$$= \left(\frac{2}{3} + \frac{1}{6} + \frac{1}{18}\right)d \times \left(\frac{2}{3} + \frac{1}{6} + \frac{1}{18}\right)d \times \left(1 + \frac{1}{2}\right)h.$$

$$= \left(\frac{2}{3} + \frac{1}{6} + \frac{1}{18}\right)\left(1 + \frac{1}{2}\right)\left(\frac{2}{3} + \frac{1}{6} + \frac{1}{18}\right)\left(1 + \frac{1}{2}\right)d \times d \times h \div \left(1 + \frac{1}{2}\right).$$

$$= \left(1 + \frac{1}{4} + \frac{1}{12}\right) \times \left(1 + \frac{1}{4} + \frac{1}{12}\right)d \times d\frac{2}{3}h.$$

$$= \left(1 + \frac{1}{3}\right)\left(1 + \frac{1}{3}\right)d \times d \times \frac{2}{3}h.$$

$$= \left(1 + \frac{1}{3}\right)d\left(1 + \frac{1}{3}\right)d\left(\frac{2}{3}h\right).$$

$$= \left(d + \frac{1}{3}d\right)\left(d + \frac{1}{3}d\right)\frac{2}{3}h. \quad \text{(khar)}$$

Problem 41 (RMP)

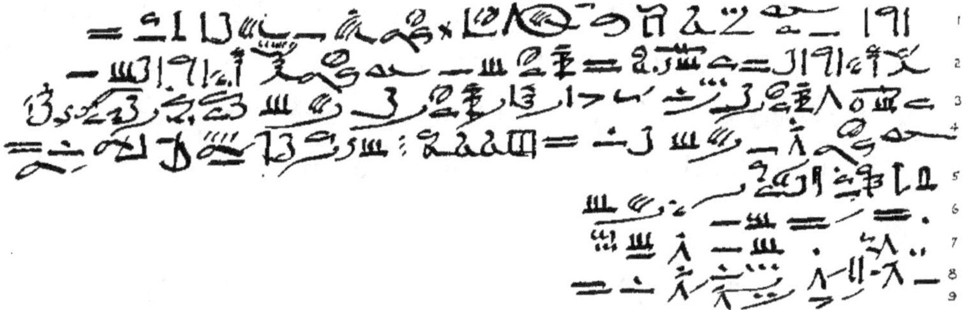

Translated as follows:

Find the volume of a cylindrical granary with a diameter of 9 cubits and a height of 10 cubits.

The Solution

The scribe used the formula below:

$$V = \left(d - \tfrac{1}{9}d\right)\left(d - \tfrac{1}{9}d\right)h$$

$$V = \left(9 - \tfrac{1}{9}\right)\left(9 - \tfrac{1}{9}\right)10 \text{ cubed cubits}$$

$$V = 640 \text{ cubed cubits}$$

$$= 640 \times \tfrac{3}{2} = 960 \text{ khar}$$

$$= 960 \times \tfrac{1}{20} = 48 \text{ hundred quadruple hekets.}$$

Problem 42 and 43 are similar to **Problem 41 (RMP).**

In the Kahun Papyrus, there are no details about the problems. All that is there are drawings accompanied by calculations which are supposed to be a cylindrical granary with a diameter of 12 cubits and a height of 8 cubits.

From the calculations, the scribe used the formula:

$$V = \left(d + \tfrac{1}{3}d\right) \times \left(d + \tfrac{1}{3}d\right) \times \tfrac{2}{3}h = \left(12 + \tfrac{12}{3}\right) \times \left(12 + \tfrac{12}{3}\right)\left(\tfrac{8\times2}{3}\right) = 1365\tfrac{1}{3}.$$

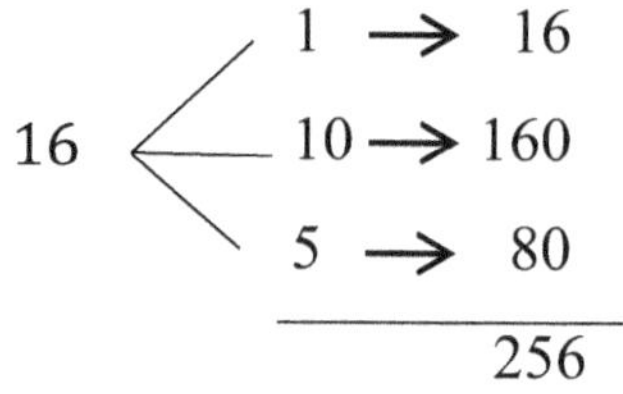

16
1 → 16
10 → 160
5 → 80
256

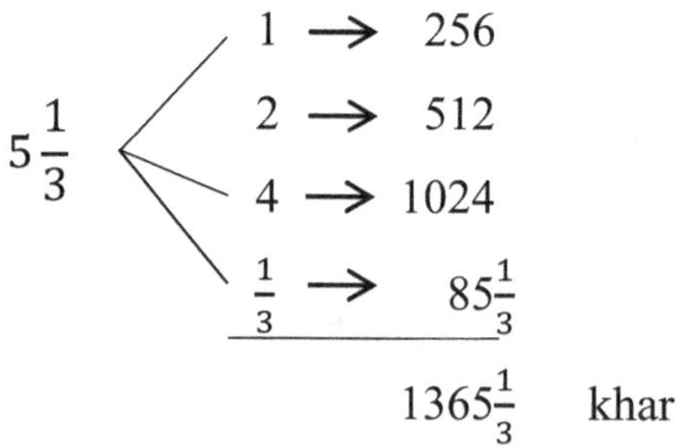

VOLUME OF THE SQUARE PYRAMIDS

Problem 40 (Cairo Papyrus)

We are asked to find the volume of a square pyramid.

Given a pyramid with a vertical height of $7\frac{1}{14}$ cubits and a square base of 10 cubits to a side, determine the volume.

The scribe used the formula $V = \frac{1}{3}bh$ where $b = s^2$ which is the area of the base and h is the height.

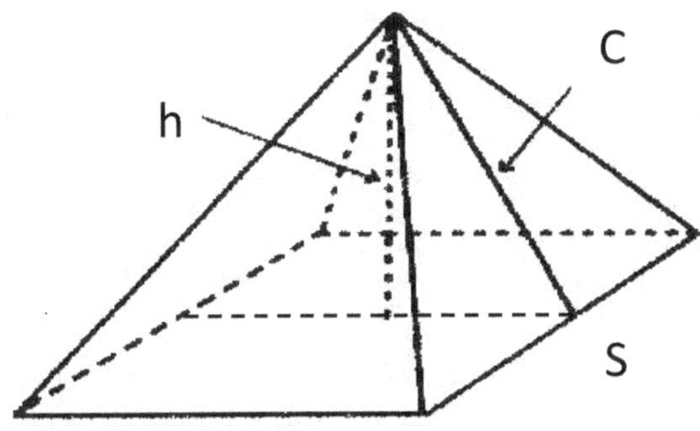

Square Pyramid

The way the ancient mathematician arrived at this formula may be by building six equal pyramids with the height equal to half of the side of the square to form a cube with the side equal to the side of the square. See the following diagram.

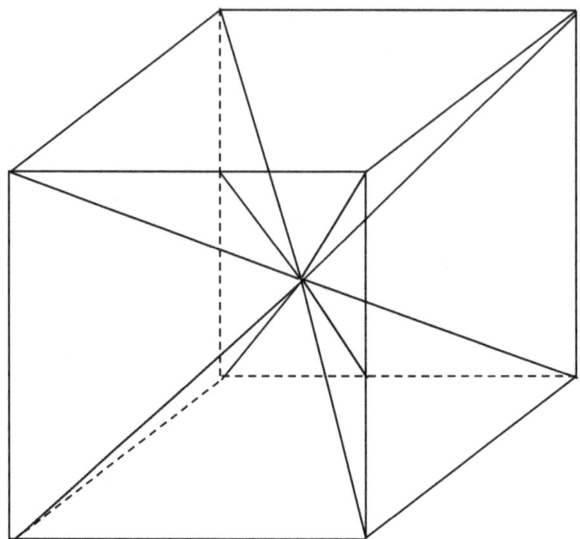

The only intuitive idea to find the volume of the pyramid is to construct a cube using 6 pyramids with each pyramid's base equal to the base of the cube and twice the height of the pyramid = the height of the cube.

The volume of the cube $= a^3$ since $a = 2h$

There are 6 pyramids in the cube.

The volume of the pyramid $= \dfrac{a^3}{6} = \dfrac{a^2 2h}{6} = \dfrac{1}{3}a^2 h = \dfrac{1}{3}bh$, and 'b' is the area of the base.

The Solution

$V = \dfrac{1}{3}bh$ where $b = 10 \times 10 = 100$, and $h = 10$.

$V = \dfrac{1}{3}bh = \dfrac{1}{3} \times 100 \times 7\dfrac{1}{14} = 235\dfrac{2}{3}\dfrac{1}{21}$ cubic cubits.

Note: See corrected problem on page 85.

VOLUME OF THE TRUNCATED PYRAMID

Problem 14 (Mosco Papyrus)

This problem also is of interest to present-day mathematicians.

You are given a truncated pyramid with a vertical height of 6 cubits, a square base with sides of 4 cubits on the bottom, and a square top with sides of 2 cubits each. Determine its volume.

The scribe used the formula $V = \frac{1}{3}(a^2 + ab + b^2)h$.

The scribe may have dissected the truncated pyramid into a rectangular object with $V_1 = a^2 h$

and 4 prisms (or side pieces) $V_2 = 4\left[\frac{1}{4}a(b-a)h\right]$

and four pyramids $V_3 = 4 \times \frac{1}{3}\left(\frac{b-a}{2}\right)^2 h$.

The volume of the truncated pyramid is $V = V_1 + V_2 + V_3$.

$V = a^2 h + 4 \times \frac{1}{4}a(b-a)h + \frac{4}{3}\left(\frac{b-a}{2}\right)^2 h$.

$\quad = a^2 h + abh - a^2 h + \frac{4}{12}(b^2 h - 2abh + a^2 h)$.

$\quad = abh + \frac{1}{3}(b^2 h - 2abh + a^2 h)$.

$\quad = \frac{1}{3}(3abh + b^2 h - 2abh + a^2 h)$.

$\quad = \frac{1}{3}(a^2 + ab + b^2)h$.

<u>The Solution</u>

$V = \frac{1}{3}(a^2 + ab + b^2)h$

$\quad = \frac{1}{3}(2 \times 2 + 2 \times 4 + 4 \times 4)6 = 56$ cubic cubits. $h = 6$

This is the correct answer. Nowhere was the formula written out; but you can see it was known to the ancient Egyptian.

If b = 0, the formula of the volume of the pyramid is $V = \frac{1}{3}a^2 h$.

One of the modern approaches to finding the volume of the truncated pyramid with base a^2 and height $h_2 (V_T)$ is to find the volume of pyramids with a base a^2 and height $h = h_1 + h_2$, then subtract the volume of the top pyramid with the base b^2 and the height h_1.

$$V_T = \frac{1}{3}a^2 h - \frac{1}{3}b^2 h_1 = \frac{1}{3}a^2(h_1 + h_2) - \frac{1}{3}b^2 h_1 = \frac{1}{3}a^2 h_1 + \frac{1}{3}a^2 h_2 - \frac{1}{3}b^2 h_1.$$

And, since $\dfrac{h_1}{h} = \dfrac{b}{a} \rightarrow \dfrac{h_1}{h_1 + h_2} = \dfrac{b}{a}$; $\therefore ah_1 = b\,h_2 \rightarrow a\,h_1 - b\,h_1 = b\,h_2$.

$$(a - b)h_1 = b\,h_2.$$

$$h_1 = \frac{bh_2}{a-b}$$

$$V_T = \frac{1}{3}a^2\left(\frac{bh_2}{a-b}\right) + \frac{1}{3}a^2 h_2 - \frac{1}{3}b^2\left(\frac{bh_2}{a-b}\right)$$

$$= \frac{1}{3}h_2\left(\frac{a^2 b}{a-b} + a^2 - \frac{b^3}{a-b}\right)$$

$$= \frac{1}{3}h_2\left(\frac{a^2 b + a^3 - a^2 b - b^3}{a-b}\right) = \frac{1}{3}h_2\left(\frac{a^3 - b^3}{a-b}\right)$$

$$= \frac{1}{3}h_2\left(\frac{(a-b)(a^2 + ab + b^2)}{a-b}\right) = \frac{1}{3}h_2(a^2 + ab + b^2)$$

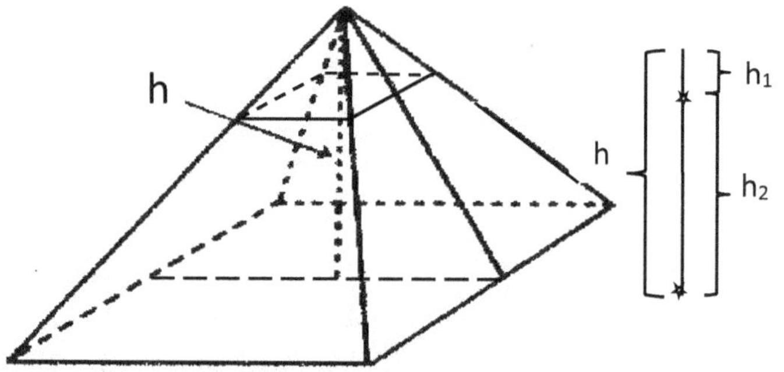

Square Pyramid

Chapter 11

Trigonometry

SEKED DEFINITION AND PROBLEMS

Trigonometry is the branch of mathematics that involves the ratios between the sides of a right triangle with reference to the acute angle (trigonometric function). It also addresses the relationship between these ratios and the application of these facts in finding the unknown side, or angle, of any right triangle.

In **Problems 56-70** in the Ahmes Papyrus, the ancient Egyptian introduced one of these ratios, which was named the seked. The seked is used to determine the height of the pyramid or to find the inclination of one of the four triangular faces to the horizontal plane of its base.

$$\text{The Seked} = \frac{(\overline{AB}),\ \text{the horizontal side of the triangle in palms}}{(\overline{BC}),\ \text{the vertical side of the triangle in cubits}}$$

This formula is interpreted as the horizontal departure of an oblique line from the vertical axis for every unit change in the height.

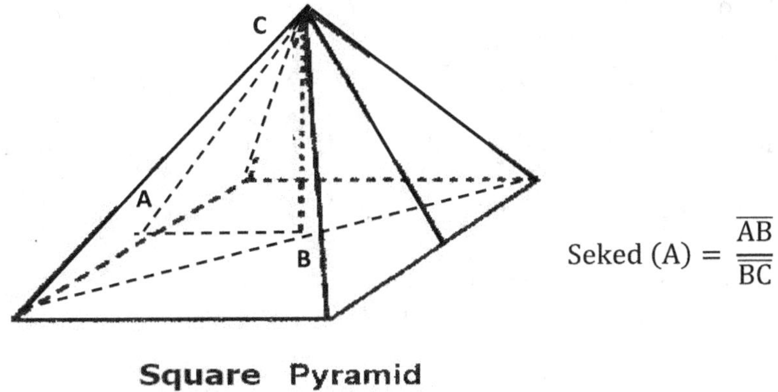

Square Pyramid

The horizontal unit of length ($\overline{AB}$) is in palms and the vertical unit of length ($\overline{BC}$) is measured by cubits.

Royal Cubit $= 7$ palms

Palm $= 4$ fingers

The seked of ($\widehat{A}$) is equivalent to the reciprocal of the tangent of $\widehat{A}$ multiplied by 7, or the cotangent of $\widehat{A}$ multiplied by 7, or the reciprocal of the slope of $\overline{AC}$ multiplied by 7 if $\overline{AB}$ and $\overline{BC}$ are in cubits.

If $\overline{BC} = 364$ cubits and $\overline{AB} = 286$ cubits,

$$\text{Tangent } \widehat{A} = \frac{\overline{BC}}{\overline{AB}} = \frac{364}{286} = \frac{14}{11} = 1.2727.$$

$$\text{Cotangent of } \widehat{A} = \frac{1}{\tan \widehat{A}} = \frac{11}{14}; \quad \text{Seked A} = \frac{11}{14} \times 7 = 5\frac{1}{2} \text{ palms per cubit.}$$

$$\text{The slope of } \overline{AC} = \tan \widehat{A} = \frac{14}{11}.$$

The reciprocal of the slope $= \frac{11}{14}$ and the seked $= \frac{11}{14} \times 7 = 5\frac{1}{2}$, which is the seked of the slope of the side of the great pyramid, and

$$\widehat{A} = 51°.84 \text{ or } 51°50'24".$$

Another famous pyramid, the Khufu Pyramid, has a seked of $5\frac{1}{4}$ and $\widehat{A} = 53°10'$. The Menkaure Pyramid has a seked $5\frac{3}{5}$ and

$$\widehat{A} = 51°.84 \text{ or } 51°50'25".$$

Most of the Egyptian pyramids have a seked between 5 and $5\frac{3}{5}$.

In the case of a pyramid with $\overline{AB} = \frac{1}{2}$ the side of the pyramid base, and $\overline{BC}$ is the height of the pyramid, and $\overline{AB}$ and $\overline{BC}$ are in cubits, then the seked $= \frac{\overline{AB}}{\overline{BC}} \times 7$ palms per cubit.

The height (altitude) of the pyramid $\overline{BC} = \dfrac{\overline{AB}\times 7}{\text{seked}} = \dfrac{\text{the base}\times 7}{2\ \text{seked}}$

$= \dfrac{7}{2\ \text{seked}}\text{(base)}$ and $\overline{AB} = \dfrac{1}{2}$ of the side of the base of the pyramid $= \overline{AB}$

$= \dfrac{\text{seked}\times \overline{BC}}{7}$

If $\overline{BC}$ is in palms, and $\overline{AB}$ is in cubits, then the seked $= \dfrac{\overline{AB}}{\overline{BC}}$

and $\overline{AB} = \text{seked} \times \overline{BC}$ in palms and $\overline{BC} = \dfrac{\overline{AB}}{\text{seked}}$ in cubits.

The following are some of the problems from the **Rhind Mathematical Papyrus (RMP)** regarding the seked. In all of these problems, the scribe drew the pyramid beside the problem, as shown.

Problem 56 (RMP)

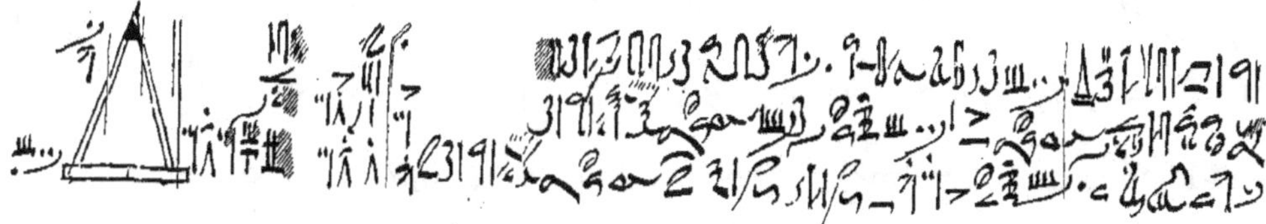

Translated as follows:

If a pyramid is 250 cubits high and the side of its base is 360 cubits long, what is the seked?

<u>The Solution</u>

Half of the side of the base $\overline{AB} = \dfrac{360}{2} = 180$ cubits and $\overline{BC} = 250$

cubits; $\dfrac{\overline{AB}}{\overline{BC}} = \dfrac{180}{250} = \dfrac{1}{2}\ \dfrac{1}{5}\ \dfrac{1}{50}.$

Seked $= 7\left(\dfrac{1}{2}\ \dfrac{1}{5}\ \dfrac{1}{50}\right) = 5\dfrac{1}{25}$ palms per cubit.

This means the slope of the triangle faces of this pyramid is $5\dfrac{1}{25}$ palms horizontally for every rise of one cubit in height.

Problem 57 (RMP)

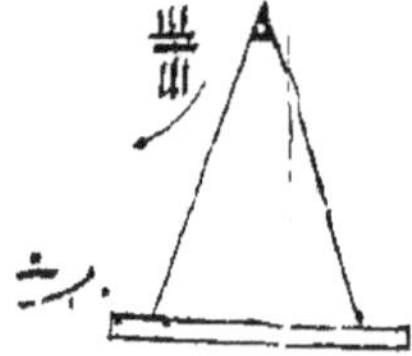 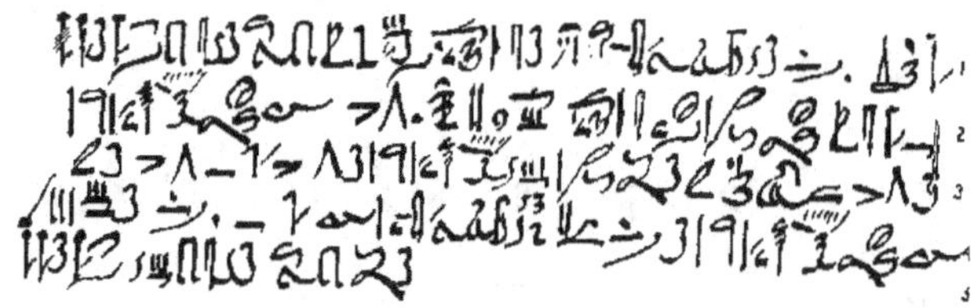

Translated as follows:

If the seked of a pyramid is 5 palms 1 finger per cubit and the sides of its base are 140 cubits, what is the altitude?

<u>The Solution</u>

The scribe used the formula: $\text{Altitude} = \dfrac{7}{2\ \text{seked}}\ (\text{base})$.

1 finger $= \dfrac{1}{4}$ palm; therefore, the seked $= 5\dfrac{1}{4}$ palms per cubit.

Divide one cubit (7 palms) by the seked doubled which is $10\dfrac{1}{2}$.

$7 \div 10\dfrac{1}{2} = \dfrac{2}{3}$

$\dfrac{2}{3} \times 140 = 93\dfrac{1}{3}$ cubits $=$ altitude or height of the pyramid.

In the **Cairo Papyrus**, the scribe used the **Pythagorean Property** to find the relationship between the three sides of the right-angle triangle in the pyramid which is the height, half of the side of the square base, and the distance from the center of any side to the apex of the pyramid. See **Problem 39** in the Cairo Papyrus.

Problem 58 (RMP)

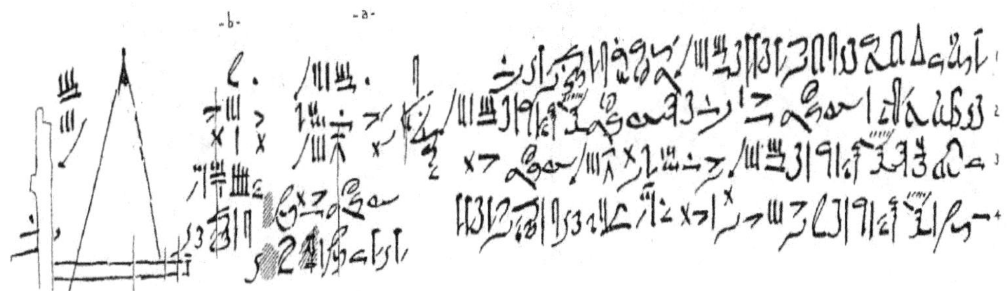

Translated as follows:

If the pyramid is $93\frac{1}{3}$ cubits high and the side of its base is 140 cubits long, what is the seked?

The Solution

The scribe takes $\frac{1}{2}$ of 140; the result is 70. Then, he divides 70 by $93\frac{1}{3}$; the result is $\frac{1}{2}\frac{1}{4}$.

$$\text{Seked} = \left(\frac{\frac{1}{2}\text{base}}{\text{height}}\right)7 = \left(\frac{70}{93\frac{1}{3}}\right)7 = \left(\frac{1}{2}\frac{1}{4}\right)(7).$$

Then, he multiplied $\frac{1}{2}\frac{1}{4}$ by 7. The result is $3\frac{1}{2} + 1\frac{1}{2} + \frac{1}{4} = 5\frac{1}{4}$ palms per cubit.

Since a finger $= \frac{1}{4}$ palm, the result is five palms and one finger per cubit.

Problem 59 and 59B

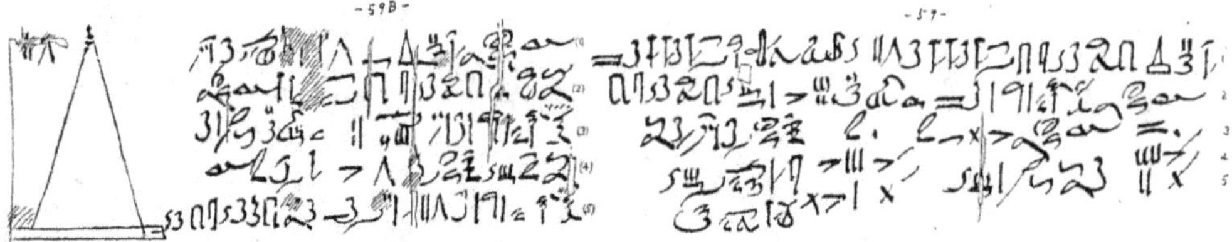

Problem 59 (RMP)

Translated as follows:

If a pyramid is 8 cubits high and has a base 12 cubits long, what is the seked?

The scribe applied the formula seked $\left(\dfrac{\frac{1}{2}\,\text{base}}{\text{height}}\right) 7$.

He found $\frac{1}{2}$ base $= \left(\frac{1}{2}\right) 12 = 6$.

He then divides 6 by 8, and the result is $\frac{1}{2}\frac{1}{4}$. Then he multiplies $\frac{1}{2}\frac{1}{4}$ by 7, and the result is 5 palms and 1 finger per cubit, which is the seked.

Problem 59B (RMP)

If the seked of a pyramid is 5 palms and one finger per cubit and the side of its base is 12 cubits long, what is its altitude?

$$\text{Seked} = \frac{\frac{1}{2}\,\text{side base}}{\text{altitude}}.$$

$$\text{Altitude} = \left(\frac{\text{side base}}{2\,\text{seked}}\right) 7 = \frac{7}{2\,\text{seked}}(\text{side base}) = \left(\frac{7}{10\frac{1}{2}}\right)(12) = 8 \text{ cubits}.$$

The scribe multiplied the seked by 2, and the result is $10\frac{1}{2}$.

Then, he divided 7 by $10\frac{1}{2} = \frac{2}{3}$.

To get the altitude, he multiplied 12 by $\frac{2}{3}$, and the result is 8 cubits.

Problem 60 (RMP)

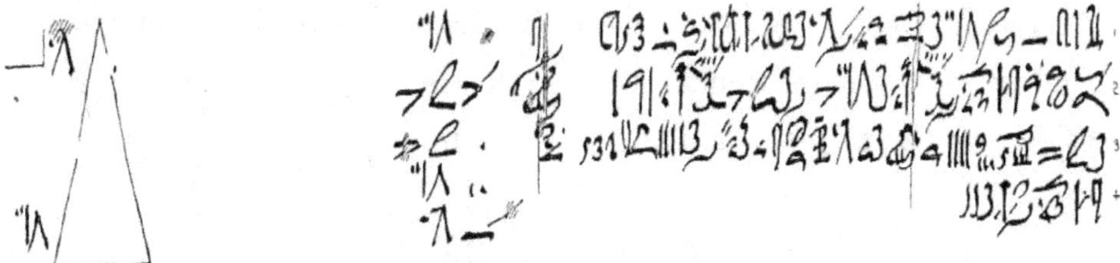

Translated as follows:

If a pillar is 30 cubits high and the side (diameter) of its base is 15 cubits, what is its seked?

The Solution

The scribe takes $\frac{1}{2}$ of 15, and the result is $7\frac{1}{2}$.

Then, he divides $7\frac{1}{2}$ by 30, and the result is $\frac{1}{4}$ cubit. (He did not multiply by 7 to express the result by palms per cubit). To complete that, we multiply $\frac{1}{4}$ x 7 which equals the seked which equals $1\frac{1}{2}\frac{1}{4}$ palms per cubit which equals 1 palm and 3 fingers per cubit. The palm is 4 fingers.

In general, the seked is used as a fraction given as so many palms horizontally for each cubit vertically, where 7 palms is equal to one cubit. The Egyptian word seked is related to our modern word "gradient." The gradient is the rate of inclination; hence, the seked of the face of the pyramid is the ratio of the run to the rise.

SHADOWS AND THE SHADOW TABLE

In **Problems 57, 59B, and 60** of the RMP, the object was to find the height of the pyramid or the pillar by using the seked. This is the first recorded equivalent trigonometric function to be used to find the height of a pyramid. The Greeks (Thales) used geometry to find the height of a pyramid, but not trigonometry as the ancient Egyptians did.

The measure of the height of a pillar ($\overline{BC}$) by means of its shadow ($\overline{AB}$) using a stick ($\overline{ED}$) and its shadow ($\overline{AE}$) requires the knowledge in the following figure:

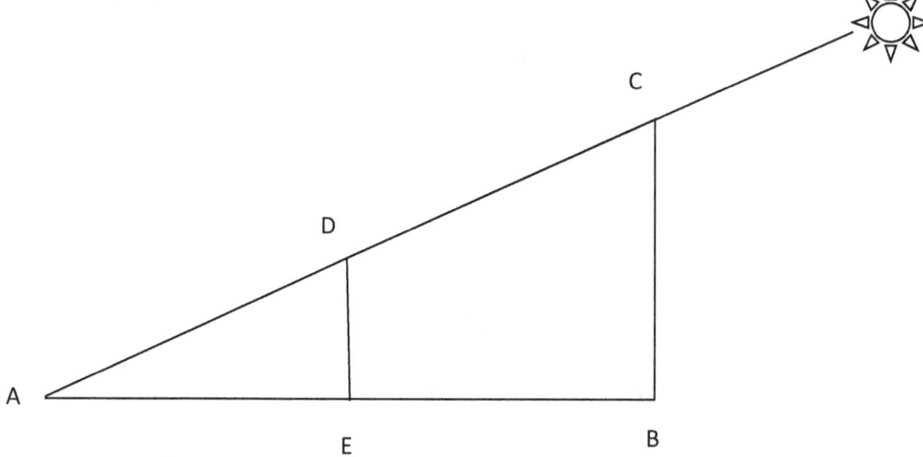

and the knowledge of the values of $\overline{DE}$, $\overline{AE}$, and $\overline{AB}$.

Using values in cubits, if $\overline{AE} = 10$ and $\overline{DE} = 14$ and $\overline{AB} = 20$,

Then the seked of $\widehat{A} = \left(\dfrac{\overline{AE}}{\overline{DE}}\right) 7 = \dfrac{10}{14} \times 7 = \dfrac{70}{14} = 5$ palms per cubit.

$\overline{AB} = 7\left(\dfrac{\overline{AB}}{\text{Seked } \widehat{A}}\right).$ $\overline{BC} = 7\left(\dfrac{20}{5}\right) = 28$ cubits.

If $\overline{BC} = 28$ cubits and $\overline{AB} = \dfrac{\overline{BC} \times \text{seked } \widehat{A}.}{7} = \dfrac{28 \times 5}{7} = 20$ cubits; then $\overline{AB}$ is the shadow of line $\overline{BC}$.

The relationship between the vertical stick (gnomon) and its shadow is the foundation of sun dials which the Egyptian used as early as 1500 BC.

Many researchers argue that the ancient Egyptian was the first time keeper around 3500 BC. They erected obelisks (tall four-sided monuments), in specific places in order to cast shadows as the sun moved overhead. They thus had a crude form of sundial which was broken into two parts; before noon and after noon. Eventually more divisions would be added, thus breaking down the time units into hours. Based on the length of the obelisk's shadow, the huge sundials could also be used to determine the longest and shortest time of the year.

The higher elevation of the sun gave a shorter shadow (essentially the seked concept). If we know the seked A_i of the end of each hour (i) and the length of the stick, we can determine the length of the shadow for each end of the hour of the day. The shadow of the stick at the end of each hour where

$$b_i = \frac{g \text{ seked } A_i}{7}.$$ 'g' is the gnomon.

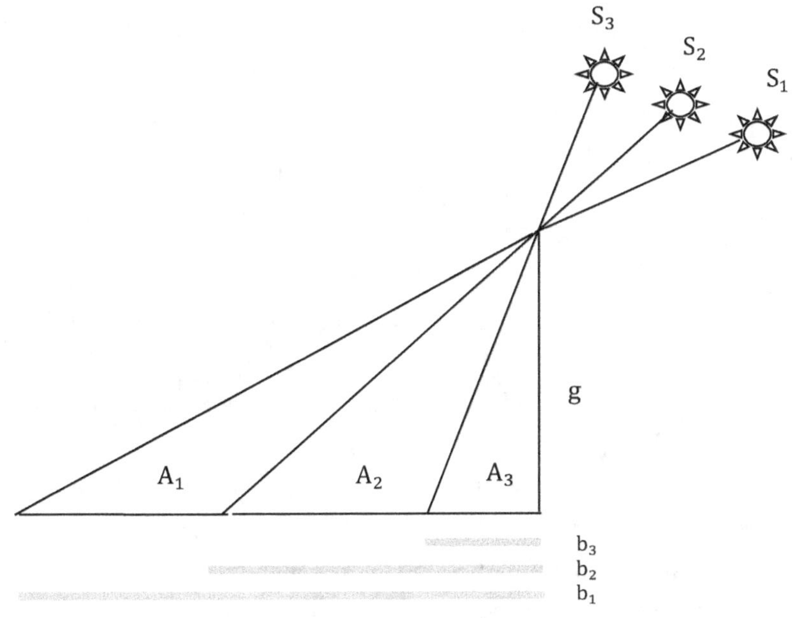

End of hr.	Shadow

2	30
3	18
4	9
5	3
Noon	

The hour of the day and the shadow of the stick (g) were recorded in an inscription from the 13th century BC found at Alydos in upper Egypt. (Neugebauer & Parker, 1960, pp. 110-127). To the left is the first recorded shadow table known to us. From the table, the Egyptians were emphasizing the idea of the function: **For any given time, there is a unique shadow length with a decreasing shadow length for consecutive increasing hours in the morning.**

The table shows a simple sequence of a pair of numbers. They are legitimate ancestors of the tangent and cotangent functions, as relates to the sundial. It is an application of the fact that the shadow of the vertical stick (gnomon) is long in the early morning and shortens to a minimum at noon.

The significance of the sundial in the photo is that it is roughly one thousand years older than what was generally accepted as time when this type of time measuring device was used, according to researcher Susanne Bickel (2008), of the University of Basel in Switzerland. (LiveScience, Science News, Jeanna Bryner, March 20, 2013.)

The sundial is made of a flattened piece of limestone, called an ostracon, with a black semicircle divided into 12 sections drawn on top. Small dots in the middle of each of the

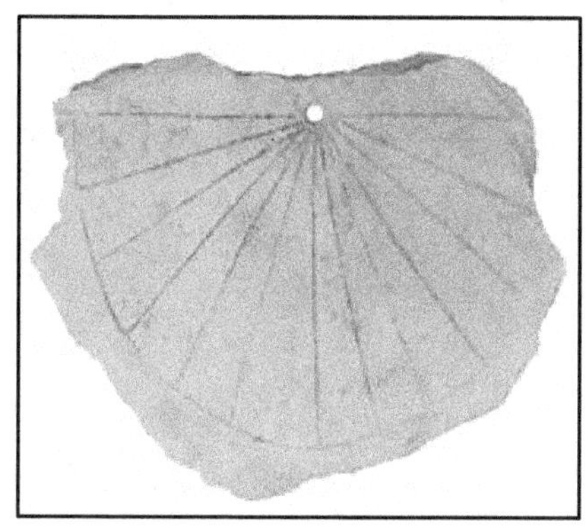

A sundial dating to the 13th Century BC was discovered in Egypt's Valley of the Kings and may be the world's oldest ancient Egyptian sundial, say scientists. Image Credit: University of Basel

12 sections, which are about 15 degrees apart, likely served to give more precise times. A dent in the center of the **ostracon** probably marks where a metal or wooden peg was inserted to cast a shadow and reveal the time of day.

From the seked problems and its applications, we can see the fundamental beginning of trigonometry and a theory of similar triangles.

Appendix I

The Contents of the Rhind (Ahmes) Papyrus

Section I: Division of 2 by 50 Odd Numbers from 3 to 101

The papyrus contains the result of the division of 2 by 50 odd numbers from 3 to 101, and the scribe proved the result using the following problems.

Section II: Division by 10

A table of the division of numbers 1-9 by 10, the result expressed in unit fractions. They proved the result using the following problems.

Problem 1: Divide one loaf among 10 men

Problem 2: Divide 2 loaves among 10 men

Problem 3: Divide 6 loaves among 10 men

Problem 4: Divide 7 loaves among 10 men

Problem 5: Divide 8 loaves among 10 men

Problem 6: Divide 9 loaves among 10 men

Section III: Multiplication of Certain of Fractional Expressions

Problem7: Multiply $\frac{1}{4}\frac{1}{28}$ by $1\frac{1}{2}\frac{1}{4}$

Problem7B: Multiply $\frac{1}{4}\frac{1}{28}$ by $1\frac{1}{2}\frac{1}{4}$

Problem 8: Multiply $\frac{1}{4}$ by $1\frac{2}{3}\frac{1}{3}$

Problem 9: Multiply $\frac{1}{2}\frac{1}{14}$ by $1\frac{1}{2}\frac{1}{4}$

Problem10: Multiply $\frac{1}{4}\frac{1}{28}$ by $1\frac{1}{2}\frac{1}{4}$

Problem11: Multiply $\frac{1}{7}$ by $1\frac{1}{2}\frac{1}{4}$

Problem 12: Multiply $\frac{1}{14}$ by $1\frac{1}{2}\frac{1}{4}$

Problem13: Multiply $\frac{1}{16}\frac{1}{112}$ by $1\frac{1}{2}\frac{1}{4}$

Problem 14: Multiply $\frac{1}{28}$ by $1\frac{1}{2}\frac{1}{4}$

Problem 15: Multiply $\frac{1}{32}\frac{1}{224}$ by $1\frac{1}{2}\frac{1}{4}$

Problem 16: Multiply $\frac{1}{2}$ by $1\frac{2}{3}\frac{1}{3}$

Problem 17: Multiply $\frac{1}{3}$ by $1\frac{2}{3}\frac{1}{3}$

Problem 18: Multiply $\frac{1}{8}$ by $1\frac{2}{3}\frac{1}{3}$

Problem 19: Multiply $\frac{1}{12}$ by $1\frac{2}{3}\frac{1}{2}$

Problem 20: Multiply $\frac{1}{24}$ by $1\frac{2}{3}\frac{1}{3}$

Section IV: The Completion

Problem 21: Complete $\frac{2}{3}\frac{1}{15}$ to be 1

Problem 22: Complete $\frac{2}{3}\frac{1}{30}$ to be 1

Problem 23: Complete $\frac{1}{4}\frac{1}{8}\frac{1}{10}\frac{1}{30}\frac{1}{45}$ to $\frac{2}{3}$

Section V: Quantity Problems

Problem 24: A quantity and its $\frac{1}{7}$ added together became 19. What is the quantity?

Problem 25: A quantity and its $\frac{1}{2}$ added together became 16. What is the quantity?

Problem 26: A quantity and its $\frac{1}{4}$ added together became 15. What is the quantity?

Problem 27: A quantity and its $\frac{1}{3}$ added together became 21. What is the quantity?

Problem 28: A quantity and its $\frac{2}{3}$ added together and from the sum $\frac{1}{3}$ of the sum subtracted, and 10 remains. What is the quantity?

Problem 29: A quantity and its $\frac{2}{3}$ are added together, and $\frac{1}{2}$ of the sum is added; then $\frac{1}{3}$ of this sum is taken and the result is 10. What is the quantity?

Section VI: Division by a Fraction Expression

Problem 30: If the scribe says, what is the quantity of which $\frac{2}{3}\frac{1}{10}$ will make 10, let him hear.

Problem 31: A quantity, its $\frac{2}{3}$, its $\frac{1}{2}$, and its $\frac{1}{7}$, added together became 13. What is the quantity?

Problem 32: A quantity, its $\frac{1}{3}$ and its $\frac{1}{4}$ added together became 2. What is the quantity?

Problem 33: A quantity, its $\frac{2}{3}$, its $\frac{1}{2}$, its $\frac{1}{7}$ added together became 37. What is the quantity?

Problem 34: A quantity, its $\frac{1}{2}$ and its $\frac{1}{4}$ added together, became 10. What is the quantity?

Section VII: Division of Hekat

Problem 35: I have gone three times into Hekat measure, my $\frac{1}{3}$ has been added to me, and I return having filled the Hekat measure. What is it that say this?

Problem 36: I have gone three times and $\frac{1}{3}$ and $\frac{1}{5}$ have been added to me. I return having filled the measure. What is the quantity that say this?

Problem 37: I have gone three times into the Hekat measure, my $\frac{1}{3}$ has been added to me, $\frac{1}{3}$ of my $\frac{1}{3}$ has been added to me, and my $\frac{1}{9}$ has been added to me. I return having filled the Hekat measure. What is it that say this?

Problem 38: I have gone three times into the Hekat measure, my $\frac{1}{7}$ has been added to me, and I return having filled the Hekat measure. What is it that says this?

Section VIII: Division of Loaves

Problem 39: Example of finding the difference of shares when 100 loaves are divided among 10 men, 50 equal shares among 6 men and 50 equal shares among 4 men. What is the difference in the shares?

Problem 40: Divide 100 loaves among 5 men in such a way that the shares received shall be an arithmetic progression and that $\frac{1}{7}$ of the sum of the largest three shares shall be equal to the sum of the smallest two. What is the difference of the shares?

Section I: Geometry

Problem 41: Find the volume of a cylindrical granary with a diameter of 9 and a height of 10.

Problem 42: Find the volume of a cylindrical granary with a diameter of 10 and a height of 10.

Problem 43: A cylindrical granary has a diameter of 9 and a height of 6. What is the amount of grain that goes into it?

Problem 44: Example of reckoning the volume of a rectangular granary. Its length being 10, its width being 10, and its height being 10. What is the amount of grain that goes into it?

Problem 45: A rectangular granary into which there has gone 7500 Quadruple Hekat of grain. What are its dimensions?

Problem 46: A rectangular granary into which there has gone 2500 Quadruple Hekat of grain. What are its dimensions?

Section II: Division of 100 Hekat

Problem 47: Suppose the scribe says to thee, let me know what the result is when 100 Quadruple Hekat are divided by 10, and its multiples, in a rectangular or circular granary.

Section III: Problems of Area

Problem 48: Compare the area of a circle and its circumscribing square.

Problem 49: Example of reckoning area. Suppose it is said to thee, What is the area of a rectangle of 10 khet by 1 khet?

Problem 50: Example of a round field with a diameter of 9 khet. What is its area?

Problem 51: Example of a triangle of land. Suppose it is said to thee What is the area of a triangle with a side of 10 khet and a base of 4 khet?

Problem 52: Example of cut-off (truncated) rectangle of land. Suppose it is said to thee, What is the area of the cut-off triangle with 20 khet for its side, 6 khet for its base, and 4 khet for its cut-off line?

Problem 53: Area of a section of a triangle. In this problem, it is not clear. The best explanation from the drawing suggests an isosceles triangle of equal sides with each side equal to 14 in. in length, and the sections were made by two lines parallel to the base.

Problem 54: What equal area should be taken from 10 fields if the sum of this area is to be seven setat?

Problem 55: What equal area should be taken from five fields if the sum of this area is 3 setat?

Section IV: Pyramid, the Relation of the Length of the Two Sides of a Triangle

Problem 56:   If a pyramid is 250 cubits high and the side of its base is 350 cubits long, what is its seked?

Problem 57:   If the seked of the pyramid is 5 palms 1 finger per cubit and the side of its base is 140 cubits, what is the altitude?

Problem 58:   If a pyramid is $93\frac{1}{3}$ cubits high and the size of its base is 140 cubits long, what is the seked?

Problem 59:   If the pyramid is 8 cubits high and the side of its base is 12 cubits long, what is its seked?

Problem 59B: If the seked of a pyramid is 5 palms 1 finger per cubit and the side of its base is 12 cubits long, what is its attitude?

Problem 60:   If a pillar is 30 cubits high and the side (diameter) of its base is 15 cubits, what is the seked?

Miscellaneous Problems

Problem 61 A: Table for multiplication of fractions.

Problem 61 B: Rule for getting $\frac{2}{3}$ of an odd number.

Problem 62:   Suppose it is said to thee, A bag containing an equal weight of gold, silver, and lead has been bought for 84 sha'ty. What is the amount in it of each precious metal, that of which is given for a deben of gold being 12 sha'ty, for a deben of silver being 6 sha'ty, and for a deben of lead being 3 sha'ty?

Problem 63:   Example of dividing 70 loaves among four men in the proportion of the number of $\frac{2}{3}, \frac{1}{2}, \frac{1}{3}$, and $\frac{1}{4}$. Let me know the share that each man receives.

Problem 64:   Suppose it is said to thee, distribute 10 hekat of barley among ten men in such a way that the shares shall be an arithmetical progression with the common difference of $\frac{1}{8}$ hekat. What is the share of each?

Problem 65:   Divide 100 loaves among 10 men, including a boatman, a foreman, and a doorkeeper who receive double portions. What is the share of each?

Problem 66:   If 10 hekat of fat are given out for a year, what is the amount used in a day?

Problem 67:   How many cattle are there in a herd when $\frac{2}{3}$ of $\frac{1}{3}$ of them make 70?

Problem 68: Suppose a scribe says to thee, four overseers have drawn 100 great quadruple hekat of grain, these gangs consisting, secretively of 12, 8, 6 and 4 men. How much does each overseer receive?

Problem 69: $3\frac{1}{2}$ hekat of meal is made into 80 loaves of bread. Let me know the amount of meal in each loaf and what is the pefsu? [pefsu (cooking ratio) the number of units of food or drink that could be made from a unit of material in the process of cooking, and it determines the relative value of any food or drink].

Problem 70: If $7\frac{1}{2}\frac{1}{4}\frac{1}{8}$ hekat of meal is made into 80 loaves of bread, what is the amount of meal in each loaf and what is the pefsu?

Problem 71: From 1 des – measure of beer, $\frac{1}{4}$ has been poured off and then the measure has been filled up with water. What is the pefsu of the distributed beer?

Problem 72: Suppose it is said to thee, 100 loaves of pefsu 10 are to be exchanged for a number of loaves of pefsu 45. How many will there be?

Problem 73: Suppose it said to thee, 100 loaves of pefsu 10 are to be exchanged for loaves of pefsu 15. How many of them will there be?

Problem 74: A thousand loaves of pefsu 5 are to be exchanged at half of the loaves of pefsu 10 and half of the loaves of pefsu 20. How many of each will there be?

Problem 75: 155 loaves of pefsu 20 are to be exchanged for loaves of pefsu 3; how many of these will there be?

Problem 76: 1000 loaves of pefsu 10 are to be exchanged for a number of loaves of pefsu 20 and the same number of pefsu 30. How many of each kind will there be?

Problem 77: Suppose it is said to thee, 10 des of beer (of pefsu 2) are to be exchanged for loaves of pefsu 5. How many loaves will there be?

Problem 78: Suppose it is said to thee, 100 loaves of pefsu 10 are to be exchanged for a quantity of beer of pefsu 2. How many des of beer will there be?

Problem 79: Sum of geometrical progression of five terms, in which the first term is 7 and the multiplier is 7 (hours, cats, mice, spelt, hekat).

Problem 80: Express the "Horus Eye fractions" in terms of hinu".

Problem 81: Express fractions of hekat as "Horus Eye" fractions in terms of hinu.

Problem 82 A, Estimation of the amount of flour made into bread.

Problem 82 B: Estimation of the daily portion of feed for geese.

Problem 83: Estimation of the feed necessary for various kinds of birds.

Problem 84:   Estimation of the feed for a stall of oxen.

Problem 85:   This is a group of hieroglyphic signs written upside down on the back of the papyrus; part of this is missing.

Problem 86:   This is some account of memorandum.

Problem 87:   A memorandum of some incidents; not very coherent.

Appendix II

The Contents of the Mosco Mathematical Papyrus (MMP)

Problem 1:   Unclear.

Problem 2:   Unreadable.

Problem 3:   Compute the length of a ship's mast given $\frac{1}{3}+\frac{1}{5}$ of the length of a cedar log originally 30 cubits long.

Problem 4:   Area of a rectangle with a base (b) = 4 and a height (h)= 10.

Problem 5:   Pefsu of a loaf of bread.

Problem 6:   Given a rectangular enclosure of 12 units and the ratio of the sides is $1:\frac{3}{4}$, find the lengths of the sides.

Problem 7:   A triangle has an area of 20 and the ratio of its height to the base is $1:2\frac{1}{2}$; find the base and the height.

Problem 8:   Exchange 100 loaves of bread with pefsu 20 with beer of pefsu 4.

Problem 9:   Pefsu of loaves of bread.

Problem 10:   Compute the area of the surface of a hemisphere.

Problem 11:   With 100 logs measuring 5 by 5, how many logs measuring 4 by 4 does this correspond to?

Problem 12:   Pefsu problem is unclear.

Problem 13:   Pefsu of loaves of bread.

Problem 14:   The base is a square with sides of 4 cubits, and the top is a square with side of 2 cubits and the height of the truncated pyramid is 6 cubits. Find the volume.

Problem 15:   Pefsu of beer.

Problem 16:   Pefsu of beer.

Problem 17:   Given a triangle with the area of 20 with a base $\frac{1}{3}\frac{1}{15}$ of its height, find the base and the height.

Problem 18:   Find the area of a length of garment-cloth measuring 5 cubits 5 palms by 2 palms.

Problem 19:   Calculate the quantity taking $1\frac{1}{2}$ times and adding to 4 to make 10.

Problem 20:   Pefsu of 1000 loaves. Horus-Eye fraction.

Problem 21:   Mixing of sacrificial bread.

Problem 22:   Pefsu loaves and beer.

Problem 23:   Find the output of a shoemaker given that he has to cut and decorate sandals (unclear).

Problem 24:   Exchange of loaves and beer.

Problem 25:   Calculate the quantity by taking 2 times and adding to itself to make 9.

Problem 14 on the volume of a truncated pyramid and Problem 10 on the area of the surface of the hemisphere are the most important problems in the history of Egyptian mathematics and according to Struve, "It became the outstanding achievement in the field of mathematics" (R. J. Gillings, 1972, page 247).

Appendix III

$$\frac{2}{n}\text{ Table from Rhind Papyrus}$$

$$\frac{2}{3} = \frac{2}{3}$$

$$\frac{2}{5} = \frac{1}{3} + \frac{1}{15}$$

$$\frac{2}{7} = \frac{1}{4} + \frac{1}{28}$$

$$\frac{2}{9} = \frac{1}{6} + \frac{1}{18}$$

$$\frac{2}{11} = \frac{1}{6} + \frac{1}{66}$$

$$\frac{2}{13} = \frac{1}{8} + \frac{1}{52} + \frac{1}{104}$$

$$\frac{2}{15} = \frac{1}{10} + \frac{1}{30}$$

$$\frac{2}{17} = \frac{1}{12} + \frac{1}{51} + \frac{1}{68}$$

$$\frac{2}{19} = \frac{1}{12} + \frac{1}{76} + \frac{1}{114}$$

$$\frac{2}{21} = \frac{1}{14} + \frac{1}{42}$$

$$\frac{2}{23} = \frac{1}{12} + \frac{1}{276}$$

$$\frac{2}{25} = \frac{1}{15} + \frac{1}{75}$$

$$\frac{2}{27} = \frac{1}{18} + \frac{1}{54}$$

$$\frac{2}{29} = \frac{1}{24} + \frac{1}{58} + \frac{1}{174} + \frac{1}{232}$$

$$\frac{2}{31} = \frac{1}{20} + \frac{1}{24} + \frac{1}{155}$$

$$\frac{2}{33} = \frac{1}{22} + \frac{1}{66}$$

$$\frac{2}{35} = \frac{1}{30} + \frac{1}{42}$$

$$\frac{2}{37} = \frac{1}{24} + \frac{1}{111} + \frac{1}{296}$$

$$\frac{2}{39} = \frac{1}{26} + \frac{1}{78}$$

$$\frac{2}{41} = \frac{1}{24} + \frac{1}{246} + \frac{1}{328}$$

$$\frac{2}{43} = \frac{1}{42} + \frac{1}{86} + \frac{1}{129} + \frac{1}{301}$$

$$\frac{2}{45} = \frac{1}{30} + \frac{1}{90}$$

$$\frac{2}{47} = \frac{1}{30} + \frac{1}{141} + \frac{1}{470}$$

$$\frac{2}{49} = \frac{1}{28} + \frac{1}{196}$$

$$\frac{2}{51} = \frac{1}{34} + \frac{1}{102}$$

$$\frac{2}{53} = \frac{1}{30} + \frac{1}{318} + \frac{1}{795}$$

$$\frac{2}{55} = \frac{1}{30} + \frac{1}{330}$$

$$\frac{2}{57} = \frac{1}{38} + \frac{1}{114}$$

$$\frac{2}{59} = \frac{1}{36} + \frac{1}{236} + \frac{1}{531}$$

$$\frac{2}{61} = \frac{1}{40} + \frac{1}{244} + \frac{1}{488} + \frac{1}{610}$$

$$\frac{2}{63} = \frac{1}{42} + \frac{1}{126}$$

$$\frac{2}{65} = \frac{1}{39} + \frac{1}{195}$$

$$\frac{2}{67} = \frac{1}{40} + \frac{1}{335} + \frac{1}{536}$$

$$\frac{2}{69} = \frac{1}{46} + \frac{1}{138}$$

$$\frac{2}{71} = \frac{1}{40} + \frac{1}{568} + \frac{1}{710}$$

$$\frac{2}{73} = \frac{1}{60} + \frac{1}{219} + \frac{1}{292} + \frac{1}{365}$$

$$\frac{2}{75} = \frac{1}{50} + \frac{1}{150}$$

$$\frac{2}{77} = \frac{1}{44} + \frac{1}{308}$$

$$\frac{2}{79} = \frac{1}{60} + \frac{1}{237} + \frac{1}{316} + \frac{1}{790}$$

$$\frac{2}{81} = \frac{1}{54} + \frac{1}{162}$$

$$\frac{2}{83} = \frac{1}{60} + \frac{1}{332} + \frac{1}{415} + \frac{1}{498}$$

$$\frac{2}{85} = \frac{1}{51} + \frac{1}{255}$$

$$\frac{2}{87} = \frac{1}{58} + \frac{1}{174}$$

$$\frac{2}{89} = \frac{1}{60} + \frac{1}{356} + \frac{1}{534} + \frac{1}{890}$$

$$\frac{2}{91} = \frac{1}{70} + \frac{1}{130}$$

$$\frac{2}{93} = \frac{1}{62} + \frac{1}{186}$$

$$\frac{2}{95} = \frac{1}{60} + \frac{1}{380} + \frac{1}{570}$$

$$\frac{2}{97} = \frac{1}{56} + \frac{1}{679} + \frac{1}{776}$$

$$\frac{2}{99} = \frac{1}{66} + \frac{1}{198}$$

$$\frac{2}{101} = \frac{1}{101} + \frac{1}{202} + \frac{1}{303} + \frac{1}{606}$$

The Analysis of the $\frac{2}{n}$ Table

The table was generated using the sequence of practical numbers. If 'n' is a practical number, then the rational number of $\frac{m}{n}$ and m < n may be represented as $\frac{\sum di}{n}$ where each di is a distinct division of 'n'. Hence, the sum provides a representation of $\frac{m}{n}$ as an Egyptian fraction.

The sequence of practical numbers is 1, 2, 4, 6, 12, 16, 18, 20, 24, 28, 30, 32, 36, 40, 42, 48, 54, 56, and 60.

1. If 'k' is a divisor of 'n', where n = kL, then $\frac{k}{n} = \frac{k}{kL} = \frac{1}{L}$.

2. If '3' is a divisor of 'n', then n = 3k. We choose '2' to be a practical number (p):

$$\frac{2\times2}{3k\times2} = \frac{4}{6k} = \frac{3}{6k} + \frac{1}{6k} = \frac{1}{2k} + \frac{1}{3k} .$$

Expanding $\frac{2}{n}$ with n $= 15, 21, 27, 33, 35, 39, 45, 51, 57, 63, 69, 75, 81, 87, 93, 99,$ we will choose practical number '2'.

 Example:

$$\frac{2}{33} = \frac{2\times2}{33\times2} = \frac{4}{66} = \frac{3}{66} + \frac{1}{66} = \frac{1}{22} + \frac{1}{66} .$$

For n = 7 and 77, we will choose practical number '4'.

 Example:

$$\frac{2}{77} = \frac{2\times4}{77\times4} = \frac{8}{308} = \frac{7}{308} + \frac{1}{308} = \frac{1}{44} + \frac{1}{308}$$

For n = 5, 11, 25, 35, 55, 65, 85, and 101, we choose practical number '6'.

 Example:

$$\frac{2}{5} = \frac{2\times6}{5\times6} = \frac{12}{30} = \frac{10}{30} + \frac{2}{30} = \frac{1}{3} + \frac{1}{15}$$

$$\frac{1}{11} = \frac{2\times6}{11\times6} = \frac{12}{66} = \frac{11}{66} + \frac{1}{66} = \frac{1}{6} + \frac{1}{66}$$

$$\frac{2}{101} = \frac{2\times6}{101\times6} = \frac{12}{606} = \frac{6}{606} + \frac{3}{606} + \frac{2}{606} + \frac{1}{606} = \frac{1}{101} + \frac{1}{202} + \frac{1}{303} + \frac{1}{606}$$

For n = 13, we choose the practical number '8'.

 Example:

$$\frac{2}{13} = \frac{2\times8}{13\times8} = \frac{16}{104} = \frac{13}{104} + \frac{2}{104} + \frac{1}{104} = \frac{1}{8} + \frac{1}{52} + \frac{1}{104}$$

For n = 17, 19, and 23, we choose the practical number '12'.

 Example:

$$\frac{2}{19} = \frac{2\times12}{19\times12} = \frac{24}{228} = \frac{19}{228} + \frac{3}{228} + \frac{2}{228} = \frac{1}{12} + \frac{1}{76} + \frac{1}{114}$$

For n = 31 and 91, we choose the practical number '20'.

Example:

$$\frac{2}{91} = \frac{2 \times 20}{91 \times 20} = \frac{40}{1820} = \frac{26}{1820} + \frac{14}{1820} = \frac{1}{70} + \frac{1}{130}$$

For n = 29, 37, and 41, we will choose the practical number '24'.

Example:

$$\frac{2}{37} = \frac{2 \times 24}{37 \times 24} = \frac{48}{888} = \frac{37}{888} + \frac{8}{888} + \frac{3}{888} = \frac{1}{24} + \frac{1}{111} + \frac{1}{296}$$

We can use the practical number '24' for all previous 'n' except when n = 31 or 91, we will get the same result.

For n = 43, we will choose the practical number '42'.

Example:

$$\frac{2}{43} = \frac{2 \times 42}{43 \times 42} = \frac{84}{1806} = \frac{43}{1806} + \frac{21}{1806} + \frac{14}{1806} + \frac{6}{1806} = \frac{1}{43} + \frac{1}{86} + \frac{1}{128} + \frac{1}{301}$$

For n = 49, we will choose the practical number '28'.

Example:

$$\frac{2}{49} = \frac{2 \times 28}{49 \times 28} = \frac{56}{1372} = \frac{49}{1372} + \frac{7}{1372} = \frac{1}{28} + \frac{1}{46}$$

For n = 47 and 53, we will choose the practical number '30'.

Example:

$$\frac{2}{53} = \frac{2 \times 30}{53 \times 30} = \frac{60}{1590} = \frac{53}{1590} + \frac{5}{1590} + \frac{2}{1590} = \frac{1}{30} + \frac{1}{318} + \frac{1}{795}$$

For n = 59, we choose the practical number '36'.

Example:

$$\frac{2}{59} = \frac{2 \times 36}{59 \times 36} = \frac{72}{2124} = \frac{59}{2124} + \frac{9}{2124} + \frac{4}{2124} = \frac{1}{36} + \frac{1}{236} + \frac{1}{531}$$

For n = 61, 67, and 71, we will choose the practical number '40'.

Example:

$$\frac{2}{67} = \frac{2 \times 40}{67 \times 40} = \frac{80}{2680} = \frac{69}{2680} + \frac{8}{2680} + \frac{5}{2680} = \frac{1}{40} + \frac{1}{335} + \frac{1}{536}$$

For n = 73, 79, 83, 89, and 95, we will choose the practical number '60'.

Example:

$$\frac{2}{83} = \frac{2 \times 60}{83 \times 60} = \frac{120}{4980} = \frac{83}{4980} + \frac{15}{4980} + \frac{12}{4980} + \frac{10}{4980} = \frac{1}{60} + \frac{1}{332} + \frac{1}{415} + \frac{1}{489}$$

For n = 97, we will choose the practical number '56'.

Example:

$$\frac{2}{97} = \frac{2 \times 56}{97 \times 56} = \frac{112}{5432} = \frac{97}{5432} + \frac{8}{5432} + \frac{7}{5432} = \frac{1}{56} + \frac{1}{679} + \frac{1}{776}$$

With these practical numbers above, we can generate $\frac{2}{n}$ in (RMP).

Sometimes we expand $\frac{2}{n}$ by a number which is not practical, such as $\frac{2}{5}, \frac{2}{25}, \frac{2}{85}$ by using 3 which is not a practical number; and also $\frac{2}{91}$ by using 10.

Example:

$$\frac{2}{91} = \frac{2 \times 10}{91 \times 10} = \frac{20}{910} = \frac{13}{910} + \frac{7}{910} = \frac{1}{70} + \frac{1}{130}$$

In finding $\frac{2}{n}$, the scribe used two methods; one to express twice the reciprocal of the odd number, and other to show what the odd number must be multiplied by to make 2. In the $\frac{2}{n}$ table, the scribe nearly always takes the second method.

However, in the case of 35, he used a third method of multiplying by the practical number of 6, which multiplied by 35 gives a result of 210. Then he multiplied 2 by 6 with a result of 12.

$$\frac{1}{35} = \frac{6}{210}$$

$$\frac{2}{35} = \frac{12}{210} = \frac{7}{210} + \frac{5}{210} = \frac{1}{30} + \frac{1}{42}$$

The Solution

35    $\frac{1}{30}$ of 35 is $1\frac{1}{6}$, $\frac{1}{42}$ of 35 is $\frac{2}{3}\frac{1}{6}$

**   6   7   5

for $\frac{1}{35}$ applied to 210 gives 6 and $2 \times 6 = 12$, or 7 and 5.

$$\frac{2}{35} = \frac{1}{30} + \frac{1}{42} \qquad \left[\frac{7}{210} + \frac{5}{210}\right]$$

The Proof

$$1 \longrightarrow 35$$

$$\frac{1}{30} \longrightarrow 1\frac{1}{6}$$

$$\frac{1}{42} \longrightarrow \frac{1}{6}$$

Note: The scribe wrote 6, 7, and 5 under the 35, $\frac{1}{30}$, and $\frac{1}{42}$ of the first line.

Since 'n' is prime such that we have $\frac{2}{7}$, Ahmes could use three methods to expand it.

He could find the value of $\frac{1}{7}$ from the table of $\frac{1}{n}$, or find $\frac{1}{7}$ by dividing 2 over 7. Then divide the denominator of the fractions by 2.

$$
\begin{array}{rcl}
1 & \longrightarrow & 7 \\
\frac{1}{2} & \longrightarrow & 3\frac{1}{2} \\
\frac{1}{4} & \longrightarrow & 1\frac{1}{2}\frac{1}{4} \\
\frac{1}{8} & \longrightarrow & \frac{1}{2}\frac{1}{4}\frac{1}{8} \\
8 & \longrightarrow & 56 \\
\frac{1}{56} & \longrightarrow & \frac{1}{8}
\end{array}
\quad 1
$$

$$\frac{1}{7} = \frac{1}{8} + \frac{1}{56} \text{ then } \frac{2}{7} = \frac{1}{4} + \frac{1}{28}$$

Or divide 2 by 7:

$$
\begin{array}{rcl}
1 & \longrightarrow & 7 \\
\frac{1}{2} & \longrightarrow & 3\frac{1}{2} \\
\frac{1}{4} & \longrightarrow & 1\frac{1}{2}\frac{1}{4} \\
4 & \longrightarrow & 28 \\
\frac{1}{28} & \longrightarrow & \frac{1}{4}
\end{array}
\quad 2
$$

then $\frac{2}{7} = \frac{1}{4} + \frac{1}{28}$

Or follow the solution method in Problem 35.

$\frac{1}{7}$ applied to 56 gives 8 (practical number 8). And $2 \times 8 = 16$ where $14 + 2$ from 56 is

$\frac{1}{4}\frac{1}{28}$ $\quad \left[\frac{14}{56} + \frac{2}{56}\right]$

Appendix IV

Egyptian Mathematical Leather Table

Another important table on the addition of fractions is the Egyptian Mathematical Leather Table.

1. $\frac{1}{8} = \frac{1}{10} + \frac{1}{40}$

2. $\frac{1}{4} = \frac{1}{5} + \frac{1}{20}$

3. $\frac{1}{3} = \frac{1}{4} + \frac{1}{12}$

4. $\frac{1}{5} = \frac{1}{10} + \frac{1}{10} *$

5. $\frac{1}{3} = \frac{1}{6} + \frac{1}{6} *$

6. $\frac{1}{2} = \frac{1}{6} + \frac{1}{6} + \frac{1}{6} *$

7. $\frac{2}{3} = \frac{1}{3} + \frac{1}{3} *$

8. $\frac{1}{8} = \frac{1}{25} + \frac{1}{15} + \frac{1}{75} + \frac{1}{200}$

9. $\frac{1}{16} = \frac{1}{50} + \frac{1}{30} + \frac{1}{150} + \frac{1}{400}$

10. $\frac{1}{6} = \frac{1}{25} + \frac{1}{50} + \frac{1}{150}$

11. $\frac{1}{6} = \frac{1}{9} + \frac{1}{18}$

12. $\frac{1}{4} = \frac{1}{7} + \frac{1}{14} + \frac{1}{28}$

13. $\frac{1}{8} = \frac{1}{12} + \frac{1}{24}$

14. $\frac{1}{7} = \frac{1}{14} + \frac{1}{21} + \frac{1}{42}$

15. $\frac{1}{9} = \frac{1}{18} + \frac{1}{27} + \frac{1}{54}$

16. $\frac{1}{11} = \frac{1}{22} + \frac{1}{33} + \frac{1}{66}$

17. $\frac{1}{13} = \frac{1}{28} + \frac{1}{49} + \frac{1}{196}$

18. $\frac{1}{15} = \frac{1}{30} + \frac{1}{45} + \frac{1}{90}$

19. $\frac{1}{16} = \frac{1}{24} + \frac{1}{48}$

20. $\frac{1}{12} = \frac{1}{18} + \frac{1}{36}$

21. $\frac{1}{14} = \frac{1}{21} + \frac{1}{42}$

22. $\frac{1}{30} = \frac{1}{45} + \frac{1}{90}$

23. $\frac{1}{20} = \frac{1}{30} + \frac{1}{60}$

24. $\frac{1}{10} = \frac{1}{15} + \frac{1}{30}$

25. $\frac{1}{32} = \frac{1}{48} + \frac{1}{96}$

26. $\frac{1}{64} = \frac{1}{96} + \frac{1}{192}$

The Analysis of the Table

For numbers 1, 2, 3, 13, 19, 20, 21, 22, 23, 24, 25, and 26 follow the rule that if :

$b = ka$; then $\frac{1}{a} + \frac{1}{b} = \frac{k+1}{b}$ providing $\frac{b}{k+1}$ is an integer

Example (#2):

$\frac{1}{5} + \frac{1}{20}$; $20 = 4 \times 5$; $k = 4$ and $k + 1 = 5$; then $\frac{1}{5} + \frac{1}{20} = \frac{5}{20} = \frac{1}{4}$

Numbers 14, 15, 16, 17, and 18 follow the rule that: $\frac{1}{a} = \frac{1}{2a} + \frac{1}{3a} + \frac{1}{6a}$

Proof: $\frac{1}{a} = \frac{1}{a} \times 1 = \frac{1}{a}\left(\frac{1}{2} + \frac{1}{3} + \frac{1}{6}\right) = \frac{1}{2a} + \frac{1}{3a} + \frac{1}{6a}$

Example (#15):

$$\frac{1}{18} + \frac{1}{24} + \frac{1}{54} \qquad 2a = 18 \Rightarrow a = 9; \text{ then } \frac{1}{2\times9} + \frac{1}{3\times9} + \frac{1}{6\times9} = \frac{1}{9}$$

If $\frac{1}{a_1} + \frac{1}{a_2} + \frac{1}{a_3}$, and $\frac{a_3}{a_1} + \frac{a_3}{a_2} + \frac{a_3}{a_3} = k$ and $a_3 = km$;

then $\frac{1}{a_1} + \frac{1}{a_2} + \frac{1}{a_3} = \frac{1}{m}$ and assuming k, m are integers

Example (#12):

$$\frac{1}{7} + \frac{1}{14} + \frac{1}{28} \text{ and } \frac{a_3}{a_1} + \frac{a_3}{a_2} + \frac{a_3}{a_3} = 4 + 2 + 1 = 7 = k$$

$a_3 = 28 = 7 \times 4 \Rightarrow m = 4 \qquad$ k, m are integers

$$\therefore \frac{1}{7} + \frac{1}{14} + \frac{1}{28} = \frac{1}{4}$$

If we apply the rule for $\frac{1}{25} + \frac{1}{50} + \frac{1}{150}$, then $\frac{a_3}{a_1} + \frac{a_3}{a_2} + \frac{a_3}{a_3} = 6 + 3 + 1 = 10 = k$

$150 = 10 \times 15 \quad$ k = 10 and m = 15; then $\frac{1}{25} + \frac{1}{50} + \frac{1}{150} = \frac{1}{15}$.

(The answer in the table is $\frac{1}{6}$ which is <u>not</u> correct.)

Also, #17 is not correct.

$$\frac{1}{28} + \frac{1}{49} + \frac{1}{196} \text{ then } \frac{a_3}{a_1} + \frac{a_3}{a_2} + \frac{a_3}{a_3} = 7 + 4 + 1 = 12 = k$$

$196 = 12(16.5) \quad$ k = 12 and m = 16.5 $\neq$ 13

The answer in the table is 13, which is <u>not</u> correct.

If we have four fractions, we can extend the rule.

$$\frac{1}{a_1} + \frac{1}{a_2} + \frac{1}{a_3} + \frac{1}{a_4}$$

If $\frac{a_4}{a_1} + \frac{a_4}{a_2} + \frac{a_4}{a_3} + \frac{a_4}{a_4} = k$ and $a_4 = km$

then $\frac{1}{a_1} + \frac{1}{a_2} + \frac{1}{a_3} + \frac{1}{a_4} = \frac{1}{m}$ and assuming k, m are integers

Example:

$$\frac{1}{25} + \frac{1}{15} + \frac{1}{75} + \frac{1}{200}$$

and $\frac{200}{25} + \frac{200}{15} + \frac{200}{75} + \frac{200}{200} = 8 + 13\frac{1}{3} + 2\frac{2}{3} + 1 = 25 = k$

and $m = \frac{200}{25} = 8$

$$\therefore \frac{1}{25} + \frac{1}{15} + \frac{1}{75} + \frac{1}{200} = \frac{1}{8}$$

The table can be broken down into four (4) groups.

Group I

The following rules will be used:

1. $\frac{1}{n} = k\sum\frac{1}{kn}$ 2. $n\left(\frac{1}{na}\right) = \frac{1}{a}$ 3. $\frac{1}{a}\left(\frac{1}{b}\right) = \frac{1}{ab}$ 4. $n\left(\frac{1}{b}\right) = \frac{1}{k}$ if $b = nk$

1. $\frac{1}{8} = 5\left(\sum\frac{1}{40}\right) = \frac{1}{40} + \frac{1}{40} + \frac{1}{40} + \frac{1}{40} + \frac{1}{40} = 4\left(\frac{1}{40}\right) + \frac{1}{40} = \frac{1}{10} + \frac{1}{40}$

2. $\frac{1}{4} = \frac{1}{5} + \frac{1}{20}$

3. $\frac{1}{3} = \frac{1}{12} + \frac{1}{12} + \frac{1}{12} + \frac{1}{12} = \left(\frac{1}{12} + \frac{1}{12} + \frac{1}{12}\right) + \frac{1}{12} = 3\left(\frac{1}{12}\right) + \frac{1}{12} = \frac{1}{4} + \frac{1}{12}$

4. $\frac{1}{5} = \frac{1}{2\times5} + \frac{1}{2\times5} = \frac{1}{10} + \frac{1}{10} \Rightarrow \frac{2}{5} = \frac{1}{5} + \frac{1}{5}$

5. $\frac{1}{3} = \frac{1}{2\times3} + \frac{1}{2\times3} = \frac{1}{6} + \frac{1}{6}$

6. $\frac{1}{2} = \frac{1}{6} + \frac{1}{6} + \frac{1}{6} = \left(\frac{1}{6} + \frac{1}{6}\right) + \frac{1}{6} = \frac{1}{3} + \frac{1}{6}$

7. $1 = \frac{1}{2} + \frac{1}{2} = \frac{1}{2} + \frac{1}{3} + \frac{1}{6} \Rightarrow \frac{1}{a}(1) = \frac{1}{a}\left(\frac{1}{2} + \frac{1}{3} + \frac{1}{6}\right) = \frac{1}{2a} + \frac{1}{3a} + \frac{1}{6a}$ *

8. $\frac{2}{3} = \frac{1}{3} + \frac{1}{3} = \left(\frac{1}{3} + \frac{1}{6}\right) + \frac{1}{6} = \frac{1}{2} + \frac{1}{6}$

Group II

If $n = 2k$, then $\frac{1}{2k} = \frac{1}{k}\left(\frac{1}{2}\right) = \frac{1}{k}\left(\frac{1}{3} + \frac{1}{6}\right) = \frac{1}{3k} + \frac{1}{6k}$

11. $\frac{1}{6} = \frac{1}{3} \times \frac{1}{2} = \frac{1}{3}\left(\frac{1}{3} + \frac{1}{6}\right) = \frac{1}{9} + \frac{1}{18}$

13. $\frac{1}{8} = \frac{1}{4} \times \frac{1}{2} = \frac{1}{4}\left(\frac{1}{3} + \frac{1}{6}\right) = \frac{1}{12} + \frac{1}{24}$

19. $\frac{1}{16} = \frac{1}{8} \times \frac{1}{2} = \frac{1}{8}\left(\frac{1}{3} + \frac{1}{6}\right) = \frac{1}{24} + \frac{1}{48}$

20. $\frac{1}{12} = \frac{1}{6} \times \frac{1}{2} = \frac{1}{6}\left(\frac{1}{3} + \frac{1}{6}\right) = \frac{1}{18} + \frac{1}{36}$

21. $\dfrac{1}{14} = \dfrac{1}{7} \times \dfrac{1}{2} = \dfrac{1}{7}\left(\dfrac{1}{3} + \dfrac{1}{6}\right) = \dfrac{1}{21} + \dfrac{1}{42}$

22. $\dfrac{1}{30} = \dfrac{1}{15} \times \dfrac{1}{2} = \dfrac{1}{15}\left(\dfrac{1}{3} + \dfrac{1}{6}\right) = \dfrac{1}{45} + \dfrac{1}{90}$

23. $\dfrac{1}{20} = \dfrac{1}{10} \times \dfrac{1}{2} = \dfrac{1}{10}\left(\dfrac{1}{3} + \dfrac{1}{6}\right) = \dfrac{1}{30} + \dfrac{1}{60}$

24. $\dfrac{1}{10} = \dfrac{1}{5} \times \dfrac{1}{2} = \dfrac{1}{5}\left(\dfrac{1}{3} + \dfrac{1}{6}\right) = \dfrac{1}{15} + \dfrac{1}{30}$

25. $\dfrac{1}{32} = \dfrac{1}{16} \times \dfrac{1}{2} = \dfrac{1}{16}\left(\dfrac{1}{3} + \dfrac{1}{6}\right) = \dfrac{1}{48} + \dfrac{1}{96}$

26. $\dfrac{1}{64} = \dfrac{1}{32} \times \dfrac{1}{2} = \dfrac{1}{32}\left(\dfrac{1}{3} + \dfrac{1}{6}\right) = \dfrac{1}{96} + \dfrac{1}{192}$

Group III

* $\dfrac{1}{a} = \dfrac{1}{a} \times 1 = \dfrac{1}{a}\left(\dfrac{1}{2} + \dfrac{1}{3} + \dfrac{1}{6}\right) = \dfrac{1}{2a} + \dfrac{1}{3a} + \dfrac{1}{6a}$

14. $\dfrac{1}{7} = \dfrac{1}{7}\left(\dfrac{1}{2} + \dfrac{1}{3} + \dfrac{1}{6}\right) = \dfrac{1}{14} + \dfrac{1}{21} + \dfrac{1}{42}$

15. $\dfrac{1}{9} = \dfrac{1}{9}\left(\dfrac{1}{2} + \dfrac{1}{3} + \dfrac{1}{6}\right) = \dfrac{1}{18} + \dfrac{1}{27} + \dfrac{1}{54}$

16. $\dfrac{1}{11} = \dfrac{1}{11}\left(\dfrac{1}{2} + \dfrac{1}{3} + \dfrac{1}{6}\right) = \dfrac{1}{22} + \dfrac{1}{33} + \dfrac{1}{66}$

18. $\dfrac{1}{15} = \dfrac{1}{15}\left(\dfrac{1}{2} + \dfrac{1}{3} + \dfrac{1}{6}\right) = \dfrac{1}{30} + \dfrac{1}{45} + \dfrac{1}{90}$

Group IV

$\dfrac{1}{a} = \dfrac{1}{a}\left(\dfrac{n}{n}\right) = n\left(\dfrac{1}{an}\right)$

12. $\dfrac{1}{4} = \dfrac{1}{4}\left(\dfrac{7}{7}\right) = 7\left(\dfrac{1}{28}\right) = \dfrac{4}{28} + \dfrac{2}{28} + \dfrac{1}{28} = \dfrac{1}{7} + \dfrac{1}{14} + \dfrac{1}{28}$

8. $\dfrac{1}{8} = \dfrac{1}{8}\left(\dfrac{25}{25}\right) = \dfrac{1}{5}\left(\dfrac{25}{40}\right) = \dfrac{1}{5}\left(\dfrac{25}{40} + \dfrac{1}{40}\right) = \dfrac{1}{5}\left(\dfrac{3}{5} + \dfrac{1}{40}\right)$

$\qquad = \dfrac{1}{5}\left(\dfrac{1}{5} + \dfrac{2}{5} + \dfrac{1}{40}\right) = \dfrac{1}{5}\left(\dfrac{1}{5} + \dfrac{1}{3} + \dfrac{1}{15} + \dfrac{1}{40}\right) = \dfrac{1}{25} + \dfrac{1}{15} + \dfrac{1}{75} + \dfrac{1}{200}$

9. $\dfrac{1}{16} = \dfrac{1}{2}\left(\dfrac{1}{8}\right) = \dfrac{1}{2}\left(\dfrac{1}{25} + \dfrac{1}{15} + \dfrac{1}{75} + \dfrac{1}{200}\right) = \dfrac{1}{50} + \dfrac{1}{30} + \dfrac{1}{150} + \dfrac{1}{400}$

We have shown that #10 and #17 are <u>not</u> correct.

APPENDIX V

Multiplication and Division Tables

Division of #1-9 by 10

$1 \rightarrow \frac{1}{10}$

$2 \rightarrow \frac{1}{5}$

$3 \rightarrow \frac{1}{5}\frac{1}{10}$

$4 \rightarrow \frac{1}{3}\frac{1}{15}$

$5 \rightarrow \frac{1}{2}$

$6 \rightarrow \frac{1}{2}\frac{1}{10}$

$7 \rightarrow \frac{2}{3}\frac{1}{30}$

$8 \rightarrow \frac{2}{3}\frac{1}{10}\frac{1}{30}$

$9 \rightarrow \frac{2}{3}\frac{1}{5}\frac{1}{30}$

The Rhind Mathematical Papyrus gives the tables in full and proves every one (1, 2, 6, 7, 8, 9)

Problem 61 (RMP)

(1) $\quad \frac{2}{3} \times \frac{2}{3} \qquad = \frac{1}{3} + \frac{1}{9}$ [3 out of 9 + 1 out of 9]

(2) $\quad \frac{1}{3} \times \frac{2}{3} \qquad = \frac{1}{2}\left(\frac{1}{3} + \frac{1}{9}\right) = \frac{1}{6} + \frac{1}{18}$

(3) $\quad \frac{2}{3} \times \frac{1}{3} \qquad = \frac{1}{6} + \frac{1}{18}$

(4) $\quad \frac{2}{3} \times \frac{1}{6} \qquad = \frac{1}{4}\left(\frac{1}{3} + \frac{1}{9}\right) = \frac{1}{2}\left(\frac{1}{6} + \frac{1}{18}\right) = \frac{1}{12}\frac{1}{36}$

(5) $\quad \frac{2}{3} \times \frac{1}{2} \qquad = \frac{1}{3}$

(6) $\quad \frac{1}{3} \times \frac{1}{2} \qquad = \frac{1}{2}\left(\frac{2}{3} \times \frac{1}{2}\right) = \frac{1}{6}$

(7) $\quad \frac{1}{6} \times \frac{1}{2} \qquad = \frac{1}{4}\left(\frac{2}{3} \times \frac{1}{2}\right) = \frac{1}{12}$

(8) $\quad \frac{1}{12} \times \frac{1}{2} \qquad = \frac{1}{8}\left(\frac{2}{3} \times \frac{1}{2}\right) = \frac{1}{24}$

(9) $\quad \frac{1}{9} \times \frac{2}{3} \qquad = \frac{1}{3}\left(\frac{1}{8} + \frac{1}{18}\right) = \frac{2}{3} \times \frac{1}{9} = \frac{1}{18} + \frac{1}{54}$

(10) $\quad \frac{1}{5} \times \frac{1}{4} \qquad = \frac{1}{20}$

(11) $\quad \frac{1}{7} \times \frac{2}{3} \qquad = \frac{1}{14} + \frac{1}{42}$

(12) $\quad \frac{1}{7} \times \frac{1}{2} \qquad = \frac{1}{14}$

(13) $\qquad \frac{1}{11} \times \frac{2}{3} \qquad = \frac{1}{22} + \frac{1}{66}$ and $\frac{1}{3} \times \frac{1}{11} = \frac{1}{33}$

(14) $\qquad \frac{1}{11} \times \frac{1}{2} \qquad = \frac{1}{22}$ and $\frac{1}{4} \times \frac{1}{11} = \frac{1}{44}$

Multiplication table for $\frac{1}{90}$ from $1 - 10$ (66 Plate 24: A9-B4 in Demotic Papyrus)

$$\frac{1}{90} = \frac{1}{90} \qquad\qquad \frac{6}{90} = \frac{1}{15}$$

$$\frac{2}{90} = \frac{1}{45} \qquad\qquad \frac{7}{90} = \frac{1}{15}\frac{1}{90}$$

$$\frac{3}{90} = \frac{1}{30} \qquad\qquad \frac{8}{90} = \frac{1}{15}\frac{1}{45}$$

$$\frac{4}{90} = \frac{1}{30}\frac{1}{90} \qquad\qquad \frac{9}{90} = \frac{1}{10}$$

$$\frac{5}{90} = \frac{1}{30}\frac{1}{45} \qquad\qquad \frac{10}{90} = \frac{1}{10}\frac{1}{90}$$

Multiplication table for $\frac{1}{150}$ from $1 - 10$ (BMP 10794)

$$\frac{1}{150} = \frac{1}{150} \qquad\qquad \frac{6}{150} = \frac{1}{30}\frac{1}{150}$$

$$\frac{2}{150} = \frac{1}{90}\frac{1}{450} \qquad\qquad \frac{7}{150} = \frac{1}{30}\frac{1}{90}\frac{1}{450}$$

$$\frac{3}{150} = \frac{1}{60}\frac{1}{300} \qquad\qquad \frac{8}{150} = \frac{1}{20}\frac{1}{300}$$

$$\frac{4}{150} = \frac{1}{45}\frac{1}{225} \qquad\qquad \frac{9}{150} = \frac{1}{30}\frac{1}{45}\frac{1}{225}$$

$$\frac{5}{150} = \frac{1}{30} \qquad\qquad \frac{10}{150} = \frac{1}{15}$$

APPendix VI

Hieroglyphics , Hieratic , And Demotic Papyri Numerals and Fractions

There are different variations in the writing of the Hieratic and the Demotic numerals and fractions. In this table one can see the similarities between the Hieratic and Demotic numerals.

Hieroglyphics	Hieratic	Demotic		Hieroglyphics	Hieratic	Demotic
1.			800.			
2.			900.			
3.			1,000.			
4.			2,000.			
5.			3,000.			
6.			4,000.			
7.			5,000.			
8.			6,000.			
9.			7,000.			
10.			8,000.			
20.			9,000.			
30.			10,000.			
40.			100,000.			
50.			$\frac{2}{3}$			
60.			$\frac{1}{2}$			
70.			$\frac{1}{4}$			
80.			$\frac{1}{8}$			
90.			$\frac{1}{16}$			
100.			$\frac{1}{32}$			
200.			$\frac{1}{64}$			
300.			$\frac{5}{6}$			
400.						
500.						
600.						
700.						

Appendix VII

The Cairo Papyrus

Recent photographs of the Cairo Mathematical Papyrus found in the Cairo Museum.

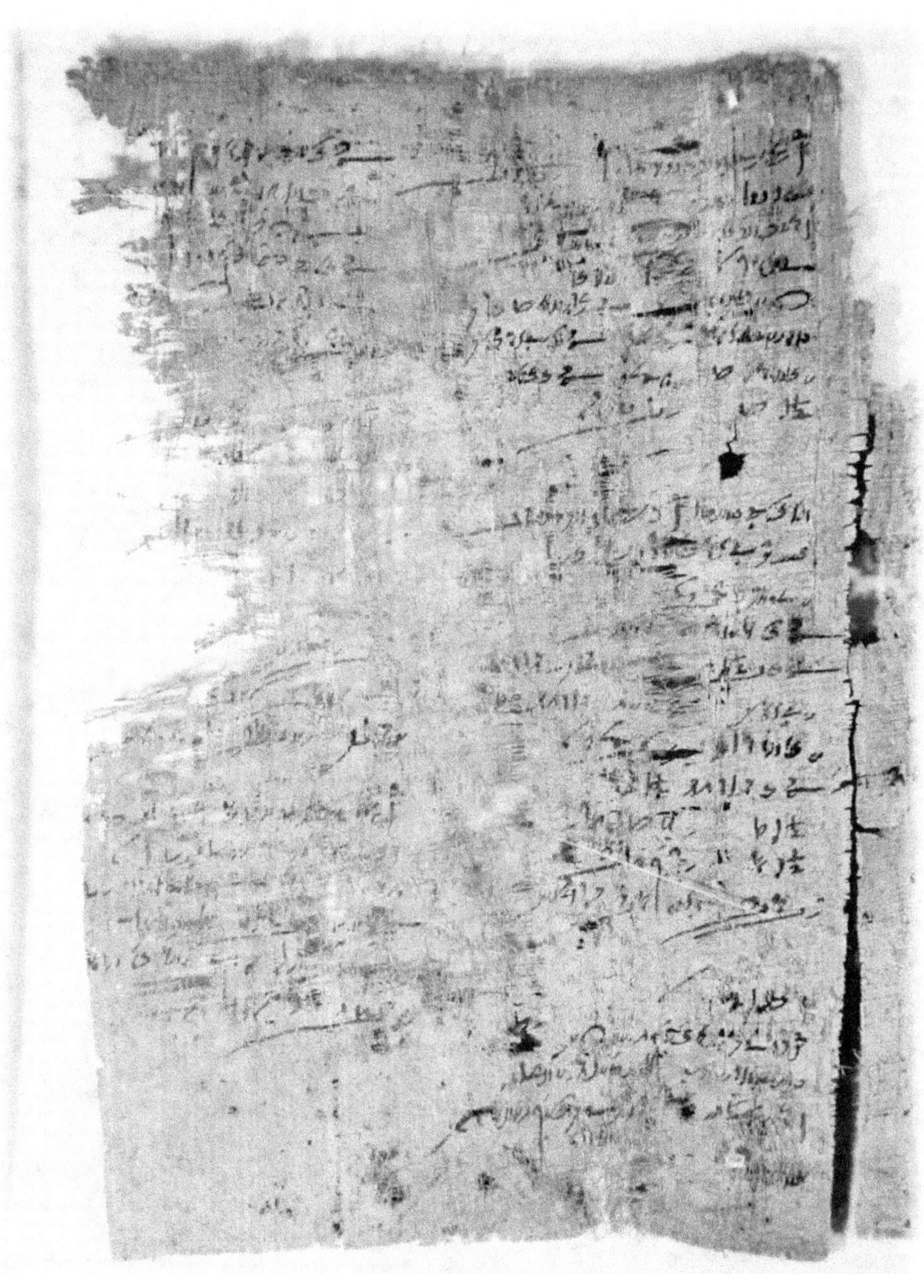

J89128

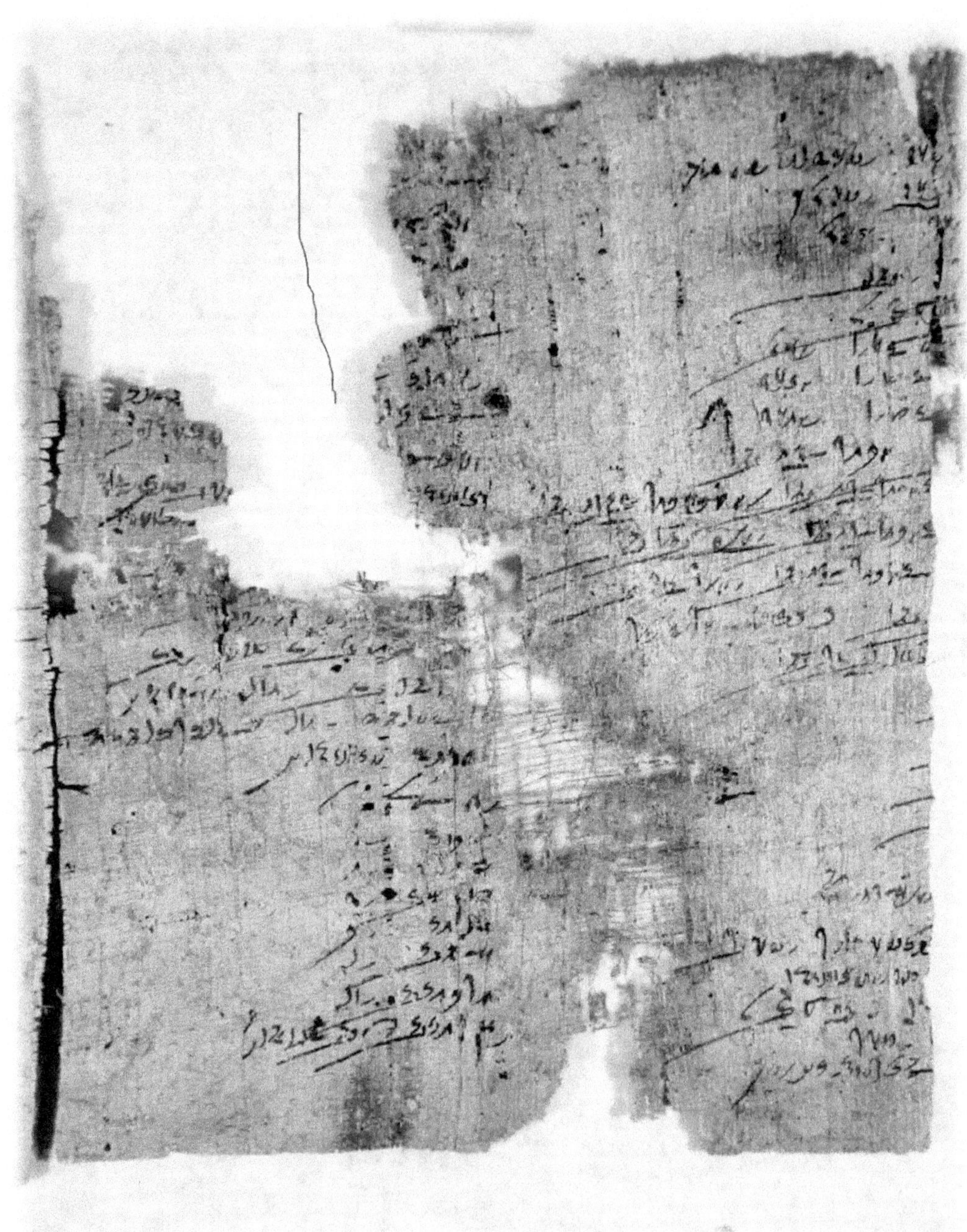

J89138

J89133

J89142
J8913O
J89127

J8913O

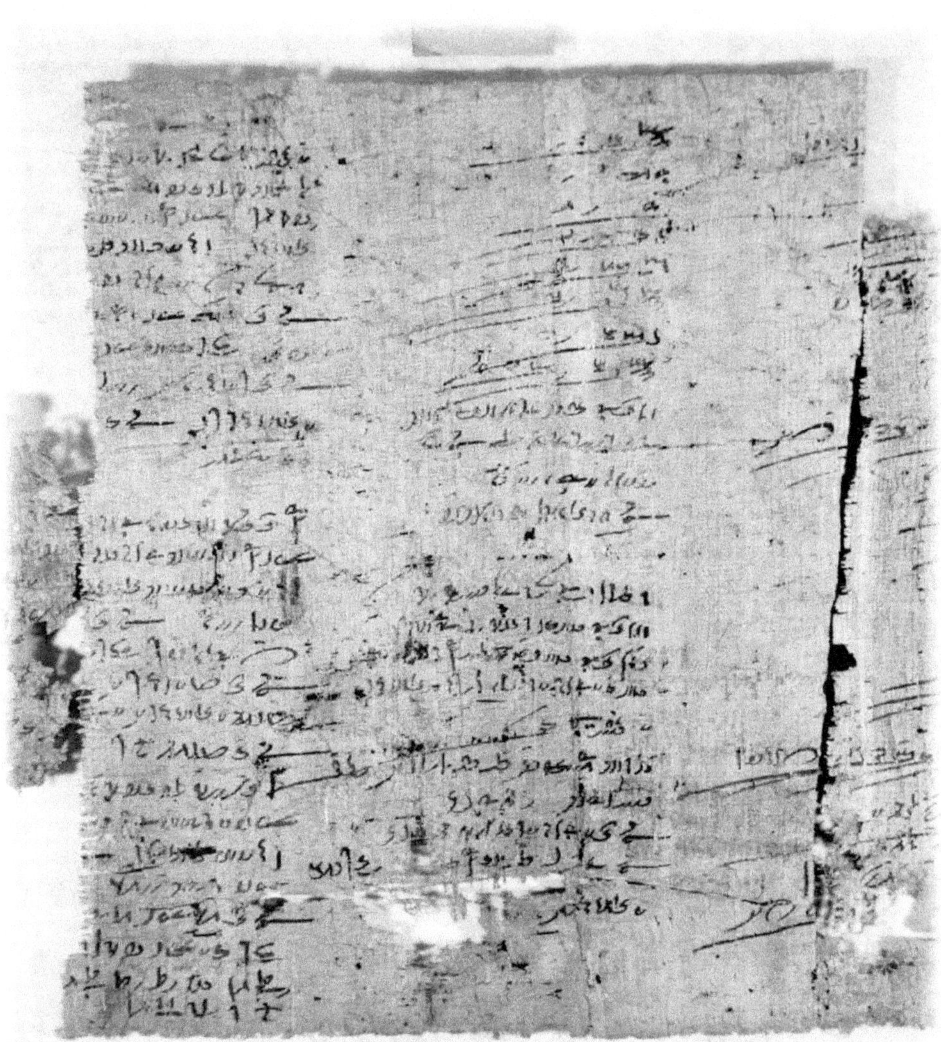

J8913O

J89142

J89143   J89141   J8914O   J89137   J89139   J89129

J89129

J89137
J89139

J8914O

J89141

J89143

REFERENCES

Alexander, T., & Thom, A. (1988). *The metrology and geometry of megalithic man*. New York, NY: Cambridge University Press.

Archibald, R. C. (1930). Mathematics before the Greeks Science. *New Series, 73*(1831), 109-121.

Barnes-Svarney, Patricia, & Svarney, Thomas E. (2006). *The handy math answer book*. Visible Ink Press.

Barton, D. (2005). *The history of mathematics: An introduction*. McGraw Hill Publishing.

Boyer, C. B., & Merzbach, Uta C. (1989). *History of mathematics,* (2nd Ed.). John Wiley & Sons.

Bruckheimer, M., & Salomon, Y. (1977). Some comments on R. J. Gillings' analysis of the $\frac{2}{n}$ table in the Rhind Papyrus. *Historia Mathematica, 4,* 445-452.

Bruins, E. M. (1981) Egyptian arithmetic. *Janus, 68,* 33-52.

Bruins, E. M. (1981). Reducible and trivial decompositions concerning Egyptian arithmetics. *Janus, 68,* 281-297.

Bruins, E. M. (1952). Ancient Egyptian arithematics. *Kon, Nederland Akardmi Van Weterschappen, Ser. A, 55*(2).

Bryner, Jeanna. Science news. *LiveScience,* March 20, 2013.

Calinger, R. (Ed.). (1995). *Classics of mathematics*. Englewood Cliffs, NJ: Prentice Hall, Inc.

Chace, A. B. (1927). *The Rhind mathematical papyrus, free translation and commentary with selected photographs, translations, transliteration and literal translation.* Classic in Mathematics 8, 2 volumes.

Chace, A. B., & Manning, H. P. (1927). *The Rhind mathematical papyrus: British Museum 10057 and 10058*. Volume 1. Mathematical Association of America, Oberlin, OH.

Clagett, M. (1999). *Ancient Egyptian science: A source book, Volume 3: Ancient Egyptian mathematics memoirs of the American philosophical society, Book 184.* Publisher: American Philosophical Society.

Clinger, R. (1999). *A conceptual history of mathematics.* Upper Straddle River, NJ.

Collection*: Egyptian Classical Near Eastern Art: Fragments of Rhind Papyrus.* Brooklyn Museum. Retrieved from http://www.brooklynmuseum.org/opencollection/objects/118304/Fragm ents_of_Rhind_Mathematical_Papyrus

Couchoud, S. (1993). *Mathématiques égyptiennes: Recherches sur les connaissances mathématiques de l'Égypte pharaonique.* Paris, France: Éditions Le Léopard d'Or

Dahan-Dalmédico, A., & Peiffer, J. (2010). *History of mathematics: Highways and byways.* Washington, DC: The Mathematical Association of America.

Daressy, G. Ostraca. *Cairo museo des antiquities Egyptiennes catalogue general ostraca hieraques, 1901,* 25001-25385.

Dunton, M., & Grim, R. E. (1966). Fibonacci on Egyptian fractions. *Fibonacci Quarterly 4,* 339-354.

Eves, H. (1990). *An introduction to the history of mathematics,* 6th Ed., Philadelphia, PA: Saunders College Publishers

Eves, H. (1969). *Historical topics for the mathematics classroom.* History of Geometry, Chapter 4, p. 169. National Council of Teachers of Mathematics, Yearbook 31.

Eves, H. (1961). *An introduction to the history of mathematics.* Austin, TX: Holt Rinehart & Winston.

Fauvel, J., & Gray, J. (1990). *The history of mathematics: A reader.* New York, NY: Macmillan Publishers.

Gardner, M. (2002). *The Egyptian mathematical leather roll: History of mathematical science.* Singapore, China: Hindustan Book Agency.

George G. M. James. (1954). Stolen legacy. *Sacred-texts.com*

Gillings, R. J. (1981). *Mathematics of the time of the pharaohs*. Mineola, NY: Dover Publications.

Gillings, R. J. (1981). The Egyptian Mathematical Leather Roll – Line 8. How did the scribe do it? *Historia Mathematica*, 456-456.

Gillings, R. J. (1974). The recto of the Rhind mathematical papyrus: How did the ancient Egyptian scribe prepare it? *Archive for History of Exact Sciences, 12*. 291-298.

Gillings, R. J. (1972). *Mathematics of the time of the pharaohs*. Cambridge: MA. MIT Press.

Gillings, R. J. (1966). Mathematics fragment from the Kahun papyrus. *Australian Journal of Science, 29*(5), 126.

Glanville, S. R. K. (1927). The mathematical leather roll in the British museum. *Journal of Egyptian Archaeology, 13,* 11, 232-283.

Griffith, F. L. (1898). *Hieratic papyri from Kahun and Gurob* (2nd. Vol.). London: GB. Bernard Quaritch, Publisher.

Gunn, B. G. (1926). Review of *The Rhind mathematical papyrus*, by T. E. Peet. *The Journal of Egyptian Archaeology, 12*, 123-137.

Hodgkin, L. (2005). *A history of mathematics from Mesopotamia to modernity*. New York, NY: Oxford University Press.

Ifrah, G. (1998). *A universal history of numbers: From prehistory to the invention of the computer*. London, England: John Wiley & Sons.

Imhausen, A. (2006). Ancient Egyptian mathematics: New perspectives on old sources. *The Mathematical Intelligencer, 28*(1), 19-27.

Imhausen, A. (2003). *Ägyptische Algorithmen*. Wiesbaden: Harrassowitz.

Imhausen, A. (2003). Egyptian mathematical texts and their contexts. *Science in Context, 16*, 367-389.

Imhausen, A. *Digitalegypt website: Lahan Papyrus IV.3*. Retrieved from http://www.digitalegypt.ucl.ac.uk/lahun/uc32160.html

Imhausen, A., & Ritter, J. (2004). Mathematical Fragments. In M. Collier & S. Quirke (Eds.), *The UCL Lahun Papyri: Religious, Literary, Legal, Mathematical and Medical* (pp. 78-79, 94-95). Oxford, Great Britain.

Imhausen, A., Robson, E., Dauben, J. W., Plofker, K., & Bergren, J. L. (2007). *The mathematics of Egypt, Mesopotamia, China, India, and Islam: A sourcebook.* V. J. Katz, (Ed.). Princeton, NJ: Princeton University Press.

James, G. G. M., 1954. *Stolen legacy.* Sacred-texts.com.. p. 43.

Johnson, G., Sriraman, B., & Saltztstein. (2012) Where are the plans? A socio-critical and architectural survey of early Egyptian mathematics. In B. Sriraman (Ed), *Crossroads in the history of mathematics and mathematics education.* Charlotte, NC: Information Age Publishing, Inc.

Joseph G. G. (1991). *The crest of the peacock: Non-European roots of mathematics* (3rd edition). pp 57-90. Princeton, NJ: Princeton University Press.

Jourdain, Philip, E. B. (1956). The nature of mathematics. In J. R. Numan (Ed), *The World of Mathematics, Vol 1.* New York, NY: Simon & Shuster.

Kline, M. (1962). *Mathematics, a cultural approach.* Boston, MA: Addison-Wesley.

Legon, J. (1992). A Kahun mathematical fragment. Based on *Discussions in Egyptology, 24,* 21-24. Retrieved from [1] http://www.legon.demon.co.uk/kahun.htm

Lumpkin, B. (2004). *The mathematical legacy of ancient Egypt: A response to Robert Palter.* National Science Foundation.

Lundsgaard, E. (1945). *Ægyptiskmatematic.* Copenhagen, Denmark

Maor, E. (1998). *Trigonometric delights.* Princeton, NJ: Princeton University Press.

Moore, D. L. (1994). *The African roots of mathematics.* (2nd ed.). Detroit, MI: Professional Educational Services

Morris, K. (1972). *Mathematical thought from ancient to modern time.* New York, NY: Oxford University Press.

Neugebauer, O. (1969)[1957]. *The exact sciences in antiquity.* Mineola, NY: Dover Publications.

Neugebauer, O., & Parker, R. A. (1960). *Egyptian astronomical texts* (Vol. 1). London, England: Brown University Press and Lund Humphries.

Newton, James. (1956). *The world of mathematics, Vol. 1*. New York: Simon & Shuster.

Oberlin: Mathematical Association of America. Reprinted Reston: National Council of Teachers of Mathematics, 1979.

Parker, R. A. (1972). *Demotic mathematical papyri*. Providence, RI: Brown University Press. London, England: Lund Humphries Press.

Peet, T. Eric. A problem in Egyptian Geometry. *Journal of Egyptian Archeology, 17*, 100-106.

Peet, T. Eric. (1923). *The Rhind mathematical papyrus, British Museum 10057 and 10058.* London: GB. The University Press of Liverpool, LTD and Hodder & Stoughton, LTD.

Rees, C. S. (1981). Egyptian fractions. *Mathematical Chronicle, 10,* 13-33.

Reimer, D. (2014). *Count like an Egyptian: A hands-on introduction to ancient mathematics*. Princeton, NJ: Princeton University Press.

Robins, G., & Shute, C. (1987). *The Rhind mathematical papyrus: An ancient Egyptian text*, London, Great Britain: British Museum Publications Limited.

Roerco, C. S. (1994). Egyptian mathematics. I. Grattan-Guinness (Ed.), *Companion Encyclopedia of the History and Philosophy of the Mathematical Sciences* (pp. 30-45). London, Great Britain.

Rossi, C. (2007). *Architecture and mathematics in Egypt*. New York, NY: Cambridge University Press.

Sarton, G. (1927). *Introduction to the history of science* (Vol 1). Baltimore, MD: Williams & Wilkins.

Scott, A., & Hall, H. R. (1927). Laboratory notes: Egyptian leather roll of the seventeenth century b.c. *British Museum Quarterly, 2,* 56.

Scott, J. F. (1969). *A history of mathematics*. London, Great Britain: Taylor and Francis.

Scott, W. (2004). *The Egyptian mathematical papayri*. Buffalo: NY. SUNY University.

Sierpiński, Waclaw. (1955). Sur une propriété des nombres naturels. *Annali di Matematica Pura ed Applicata, 39*(1), 69-74. doi: 10.2307/2372651, JSTOR 2372651, MR 0064800.

Simpson, W. K. *The Riesner Papyrus 1(1963), II(1965), III(1969)*. The Museum of Fine Art. Boston, MA.

Smith, D. E. (1958). *History of mathematics* (Vols. 1 & 2). Mineola, NY: Dover Publications.

Spalinger, A. (1990). *The Rhind mathematical papyrus as a historical document.* Hamburg, Germany: Helmut Buske Verlag GmbH

Srinivasan, A. K. (1948). Practical numbers (PDF). *Current Science, 17*, 179-180. MR 0027799

Stewart, B. M. (1954). Sums of distinct divisors. *American Journal of Mathematics, 76*(4) 779-785, doi: 10.2307/2372651, JSTOR 2372651, MR 0064800.

Strouhal, E., Vachala, B., & Vymazalová, H. (2014). *The medicine of the ancient Egyptians.* New York, NY: The American University in Cairo Press.

Struik, D. J. (1948). *A concise history of mathematics.* New York, NY: Dover Publishing.

Struve, V., & Turaev, B. (1930). *Mathematischer Papyrus des Staatlichen Museums der Schönen Künste in Moskau.* Quellen und Studien zur Geschichte der Mathematik; Abteilung A: Quellen I. Berlin: J. Springer.

van der Waerden, B. L. (1983). *Geometry and algebra in ancient civilization.* New York, NY: Springer-Verlag.

Von Baravalle, H. (1969). *Historical topics for the mathematics classroom.* The Number π, Chapter 3, p. 149. National Council of Teachers of Mathematics, Yearbook 31.

INDEX

Least common denominator (LCD): 35, 119, 120, 126, 137
Lesson book for a school boy: xii
Liber Abaci: 20, 22
Librarians: xv
Library of Congress: xiv
Limestone: 260
Limited fraction: 25
Linear units: 188-189, 190
Linguists: 86
Literary texts: 39, 86
Loaves problems, exchange using algebra: 19, 32-33, 153-161, 261, 263, 265-269
Luxor: 1

Magnitude: 220
Maja: 187
Mast: 86, 88-94
Mathematical nature: xii
Mathematical propositions and theorems: xii
Mathematical reasoning: 146
Mathematical rules and theorems: xv
Maxim of teaching: xiv
Means: 214, 218
Measurements
  area units of measurement: 190
  aroura: 108, 190-191
  cubit strip: 187, 190-191
  deben: 194, 265
  des: 155
  digit: 17, 132
  dja: 193
  double hekat: 193
  finger: 188, 189, 252, 254-257, 265
  hand: 188-189
  hekat: 2, 129, 144, 153, 155, 157-160, 196-213, 243, 263, 267
  heqat: 192-193
  heseb: 191
  hin: 86, 89, 92-94, 143, 193
  hini: 193
  hinu: 192, 193, 196-198, 201, 204, 266
  khat: 215
  khet: 190-191, 215, 217, 219, 230, 264
  kite: 194
  measuring rod: 187
  meret: 215

palm: 6, 92, 188, 189, 250-257, 265, 269
quadruple hekat: 192, 242, 244, 264, 266
remen: 191
ro: 155
rod: 187-189
royal cubit: 187-189
sack: 192
short cubit: 188
shoulder: 191
ta: 189
teper: 215
Medical knowledge: 5, 193
Medieval period / kingdom: 19, 38
Memphis, priest: 61
Menkaure Pyramid: 252
Meret: 215
Merzbach: 132, 241
Mesopotamian problems, pole: xii
Method of false position: 151, 162, 166
Method of general rules: 151
Methods of estimation: 214
Middle Ages: 136, 147
Middle Demotic: 39
Middle Kingdom: 5, 194
Midpoint: 222
Mistakes: xii, xiii, xv
Mixed number: 133
Mixed numerators: 120, 126
Modern-day mathematics: xiv
Modern mathematical analysis: 127
Modern measures: 193
Mosco Mathematical Papyrus (MMP): xv, 3, 89, 125, 151, 161, 166, 187, 214, 217, 238, 268
Moscow: xv, 3
Multiplicand: 14, 136
Multipliers: 115, 142-144
Multiplication: 150, 164
  multiplication table for $\frac{1}{150}$ from 1 to 10: 88, 109, 131, 280
  multiplication table for $\frac{1}{90}$ from 1 to 10: 88, 108, 130, 280
  multiplication table for 64 from 1 to 16: 87, 100
  multiplication table of the fractions: 130-131
  multiplication using the distributive property: 101, 114, 136, 140-141
  multiplication by 10: 113

Multiplication procedures: xii
Muses: 222
Museums: xv
Music: 153
Mysteries: xi

'n' root: 143
Nagéd Deir: 5
Natural numbers: 146, 183
Near East: 136
Neugebauer, O.: xi, 105, 260, 300
New Kingdom: 194
Newton, J.: xii
Nicomachos: 153
Nile Valley: 214
Nine Chapters of Mathematical Art: 162
19th Century: 5
Non-unit fractions: 117
Non-zero power: 23
Noon: 259, 260
Not-so-elegant Egyptian fraction: 133
Numeration: 132-133
Numerator: 21, 36, 118, 119, 126, 133, 137
Numerical; errors / mistakes: xv

Obelisk: 259
Oblique line: 251
Octagon within a square: 227-228
Odd and even numbers
   even: 25, 29, 134
   odd: 26, 27, 29, 37, 118, 127, 134, 261, 265,
   273
Official language: 39
Official script: 39
Old Kingdom: 194
Old World 146
Omission: xiv
Open mouth: 16-17, 133
Organizations: xii
Original multipliers: 14, 148
Ostracon: 260

P. British Museum: 187
P. British Museum 10399 papyrus: xiv, 86, 89,
   118
P. British Museum 10520 papyrus: xiv, 87, 98,
   114, 117, 120, 123, 124-125, 144, 182, 190

P. British Museum 10794 papyrus: xiv, 88, 108,
   130-131, 280
P. Carlsberg 30: xiv, 88, 110-111, 118-119
Paciolis, Suma: 162
Palermo Stone: 187, 190
Palm: 6, 92, 187-188, 251-258, 265, 269
Parker, R. A.: xv, 39-40, 61, 260, 300
Peet, T.: 2, 153, 238-239, 240, 298, 300
Pefsu: 3, 151, 153-161, 266-269
Pefsu definition: 153
Peg: 260
Perfect number: 23
Persian Period: 39
Petrie, W. M. F.: 7
Pharaoh Ramses II: 1
Philarch de Repugn: 61, 223
Photograph: xv
Pi (π): 41, 77-78, 80, 82, 91-94, 230, 237-239
   The Hieratic π: 227-228
   The Demotic π: 229
Pillar: 257-258, 265
Plot area: 174, 233
Pole problems: xii, 41, 61-67, 171, 223-225
Polygon: 229-230
Polygonal numbers: 99
Positional: 9-10
Positive integers: 22, 37, 143
Positive rational fraction: 132
Positive rational number: 18
Power of 10: 147
Power of 2: 23, 147
Practical application for Egyptian fraction: 18-19
Practical number: 22-24, 138, 205, 271-274
Practical problems: xii
Pregnancy testing: 5
Priests, Egyptian: xi, 214, 223
Prime denominator: 27
Prime expansion: 29
Primitive: xi
Primorial: 23
Prisms: 242, 248
Problem texts (procedural): 112
Process of completion (3 steps): 121
Professor: xv
Progression: 110, 176, 181, 182
Progression of the second order: 87, 99, 182
Proper and improper fractions: 119, 126, 133

Scaled/scaling: 8, 38, 123, 125, 201-202, 205, 207-208, 211-212
Schack-Schockenburg, Hans: 5, 7, 167
Scholars: xi, xii, xiv
Science and mathematics: xi
Scientific attitude of mind: xi
Scientific mathematics: xi
Scientific quality: xiii
Scott, Alexander: 4, 300
Second century: 86, 88
Second degree equations: 3, 5, 151
Second progression: 99-100
Seked: 251-258, 260, 265
Seked to find height of pyramid: 241-258, 265
Semi-circle: 241
Semi-cylinder: 238-240
Sequence: 152
Series notations: 18
Sessostris Pyramid: 7
Setat: 187, 190-191, 217-219, 221-222, 264
Seth: 17
Sexagesimally: 232
Shadow: 258-260
Shadow table: 260
Shaty: 194
Shematy: 194
Short cubit: 188
Shoulder: 191
Shute, C.: 2, 300
Sierpiński, Waclaw: 22, 301
Silvestre de Sacy: 86
Simpson, W. K.: 6, 301
Simultaneous equations: 3, 5, 49, 56, 151
Sinan Ibn-al-Fath: 162
Six mathematical operations: 132
Sixteenth Century: 162
Skills: xiii
Slope: 250, 251, 252
  of the triangle faces: 251
Small odd prime: 27
Smith: 162, 176, 229
Solving arithmetic progression using the average term: 176, 186
Spelt: 184-185
Square: 78-79, 83
Square cubit: 49-60, 69-82, 93, 107-108, 173, 187, 190-191, 215, 225, 232, 236-240

Square khet: 190
Square root: 64-66, 72, 73, 75, 82, 105-106, 167-170
Square root, rational: 125
Square root (square root of fractions): 57-60, 68-70, 72-73, 75, 79, 83-84, 87, 105-106, 125, 127, 143-145, 224, 231
Square pyramid: 83-85, 89, 247, 250
Squaring numbers: 143
Srinivasan, A. K.: 22, 301
Step-by-step instructions: 112, 127
Stewart, B. M.: 22, 99, 301
Stoic: 61, 223
Strouhal, Vachala, Vymazalova: 193, 301
Struik Dirk. J.: xi, 223, 301
Struve, Vasilij: 3, 125, 269, 301
Students: xi, xiv, xvi
Subtended: 77
Summands: 18
Summations: 151
Sun dials: 259-260
Sundisk: 260
Supreme: xi
Switches: 148
Symbols: 135
Switzerland: 260

Ta: 191
Tables: 2
Table of squares: 144
Table of squares of integers: 125, 127
Table of division of 2 by 50 odd numbers from 3 to 101: 2, 25, 118, 127
Table of multiplication of 64 from 1 to 16: 87, 100
Table, expanding the fractions: 24, 133
Table, finding $\frac{2}{3}$ of any fraction: 30, 112
Table of hekats in terms of hinu: 196-198
Table, Unit Fractions: 4, 133-134, 195, Appendix IV
Table, $\frac{n}{10}$: 137
Table, $\frac{2}{n}$: Appendix III: 7, 24, 27, 37-38, 113, 116, 118, 128, 137, 205, 270-273, 296
Table texts: xiii, 112, 113
Tangent: 252, 260
Teachers: xv

Temple: 1, 127, 195, 232
  chambers: 6, 127
Tenth Century: 162
Teper: 215
Text format: 127
Textbooks: xiii
Thales: xi, 222, 258
Thales Theorem: 222
Thebes: 1, 4
  priest: 61
Thebon: 1
Theodorus (of Cyrene): xi
Theon (of Smyrna): 153
Theoretical problems: xiii
Theory of arithmetic remainders: 201
Theory of arithmetic series: 99
Theory of Congruence: 214, 218
Theory of proportionality: 214
Theory of similar triangles: 260
Third century: 162
Third dimension: 243
Third rule: 143
Thirteenth century: 4, 260
Thirteenth Dynasty: 3, 5
Thoth: 17
Thoth, Cubit of: 92, 187, 189
Thrasyllus (of Mendes): xi
Three circular segments: 233
Three equations for 'x': 54, 55
Three kinds of numbers: 132
Three steps of multiplication: 141
Three steps for the process of completion: 121
Three systems of writing: xii, 132
Timekeeper: 259
Tombs: 187
Translations: xv
Trapezoid: 187, 218-219, 221, 222
Traveling: xi
Treasure: xiv
Treatises of mathematics: xii
Triangles: 187
  triangle, faces: 253
  triangle, Isosceles: 215, 220, 227
  triangle, right: 3, 41, 78, 215, 223, 251, 254
Triangular numbers: 99, 152, 182
Trigonometry: 2, 251, 258
  seked: 251-257, 260, 265

equivalent of seked: 252
height of pyramid using seked: 251-256, 265
problems using seked concept: 251-260, 265
relationship between stick's shadow and
  seked: 257-260
Truncated cone: 89
Truncated (pyramid): 3, 89, 249-250
Tuna-El-Gabel: 40
Tunis (in Fayum): 88
Turin Museum: 187
Twelfth Dynasty: 3, 5, 6
25th Dynasty: 39
26th Dynasty: 39
27th Dynasty: 39
Two thirds of any fraction: 30, 112

UC 32159: 7
UC 32160: 7
Unit fractions: 2, 4, 16, 25, 31, 87, 98, 101, 126,
  133, 134, 137, 138, 143, 195
Unit fractions, addition: 4
Unit fractions, expansion: 4, 20-22, 24-26
Unit of volume and capacity: 192-193
Unit numerators: 30
University College of London: xiv, 7
University of Basel: 260
University of Copenhagen: 88
University of Pennsylvania (Philadelphia): xiv
University of South Wales: 25
Upper Egypt: 1

van Cemlen, Luddph: 229
van der Warren, B. L.: 223
Vertical axis: 251
Viete, Francois: 229
Visiting: xi
Volume: 2, 86, 84-85, 92-93
  of cylinders: 7, 192, 214, 240, 241, 242,
    263
  of frustum/cone: 3, 89-91
  of frustum of the square pyramids: 89
  of prisms: 242
  of rectangular granary: 143, 192, 242
  of rectangular solids: 214, 242, 244
  of square pyramids: 83-85, 192, 214, 242,
    247, 248, 250
  of truncated pyramid: 214, 249-250

www.ingramcontent.com/pod-product-compliance
Lightning Source LLC
Chambersburg PA
CBHW080858160726
48000CB00009B/2773